全国高等中医药院校中药学类专业双语规划教材
Bilingual Planned Textbooks for Chinese Materia Medica Majors in TCM Colleges and Universities

有机化学实验与指导

Organic Chemistry Experiment and Guidance

（供中药学类、药学类及相关专业使用）
(For Chinese Materia Medica, Pharmacy and other related majors)

主　编　赵　骏　胡冬华

副主编　韩　波　杨　静　张园园　徐春蕾　盛文兵

编　者（以姓氏笔画为序）

方玉宇（成都中医药大学）　尹　飞（天津中医药大学）
刘晓芳（山西中医药大学）　杨　静（河南中医药大学）
李贺敏（南京中医药大学）　李嘉鹏（天津中医药大学）
肖新生（湖南科技学院）　张　薇（北京中医药大学）
张园园（北京中医药大学）　张艳春（安徽中医药大学）
林玉萍（云南中医药大学）　赵　骏（天津中医药大学）
胡冬华（长春中医药大学）　施小宁（甘肃中医药大学）
姚惠文（湖北中医药大学）　桂清文（湖南农业大学）
贾鹏昊（天津中医药大学）　徐秀玲（浙江中医药大学）
徐春蕾（南京中医药大学）　郭占京（广西中医药大学）
盛文兵（湖南中医药大学）　韩　波（成都中医药大学）

中国健康传媒集团
中国医药科技出版社

内容提要

《有机化学实验与指导》是“全国高等中医药院校中药学类专业双语规划教材”之一。本书立足于双语教学实际需要，以实验技能训练为主。全书内容分为四部分：有机化学实验的一般知识、基本操作、有机化合物合成实验、天然有机物提取纯化及简单药物合成。内容编写采用先英文，后中文对照，力求兼顾两种语言的表达特点。本教材为书网融合教材，即纸质教材有机融合电子教材，教学配套资源（PPT、微课、视频）等，使教学内容更加立体、生动、形象，便教易学。

本书适合高等中医药院校中药学类、药学类及相关专业师生、实验技术人员等学习使用，也可供从事中药研究、生产、销售工作的人员参考。

图书在版编目（CIP）数据

有机化学实验与指导：汉英对照 / 赵骏，胡冬华主编 .—北京：中国医药科技出版社，2020.8

全国高等中医药院校中药学类专业双语规划教材

ISBN 978-7-5214-1889-7

Ⅰ.①有… Ⅱ.①赵… ②胡… Ⅲ.①有机化学 – 化学实验与指导 – 双语教学 – 中医学院 – 教材 – 汉、英 Ⅳ.①O62-33

中国版本图书馆 CIP 数据核字（2020）第 100411 号

美术编辑 陈君杞

版式设计 辰轩文化

出版 **中国健康传媒集团** | 中国医药科技出版社

地址 北京市海淀区文慧园北路甲 22 号

邮编 100082

电话 发行：010-62227427 邮购：010-62236938

网址 www. cmstp. com

规格 889 × 1194 mm 1/16

印张 15

字数 381 千字

版次 2020 年 8 月第 1 版

印次 2020 年 8 月第 1 次印刷

印刷 三河市万龙印装有限公司

经销 全国各地新华书店

书号 ISBN 978-7-5214-1889-7

定价 52.00 元

出版说明

近些年随着世界范围的中医药热潮的涌动，来中国学习中医药学的留学生逐年增多，走出国门的中医药学人才也在增加。为了适应中医药国际交流与合作的需要，加快中医药国际化进程，提高来中国留学生和国际班学生的教学质量，满足双语教学的需要和中医药对外交流需求，培养优秀的国际化中医药人才，进一步推动中医药国际化进程，根据教育部、国家中医药管理局、国家药品监督管理局等部门的有关精神，在本套教材建设指导委员会主任委员成都中医药大学彭成教授等专家的指导和顶层设计下，中国医药科技出版社组织全国50余所高等中医药院校及附属医疗机构约420名专家、教师精心编撰了全国高等中医药院校中药学类专业双语规划教材，该套教材即将付梓出版。

本套教材共计23门，主要供全国高等中医药院校中药学类专业教学使用。本套教材定位清晰、特色鲜明，主要体现在以下方面。

一、立足双语教学实际，培养复合应用型人才

本套教材以高校双语教学课程建设要求为依据，以满足国内医药院校开展留学生教学和双语教学的需求为目标，突出中医药文化特色鲜明、中医药专业术语规范的特点，注重培养中医药技能、反映中医药传承和现代研究成果，旨在优化教育质量，培养优秀的国际化中医药人才，推进中医药对外交流。

本套教材建设围绕目前中医药院校本科教育教学改革方向对教材体系进行科学规划、合理设计，坚持以培养创新型和复合型人才为宗旨，以社会需求为导向，以培养适应中药开发、利用、管理、服务等各个领域需求的高素质应用型人才为目标的教材建设思路与原则。

二、遵循教材编写规律，整体优化，紧跟学科发展步伐

本套教材的编写遵循“三基、五性、三特定”的教材编写规律；以“必需、够用”为度；坚持与时俱进，注意吸收新技术和新方法，适当拓展知识面，为学生后续发展奠定必要的基础。实验教材密切结合主干教材内容，体现理实一体，注重培养学生实践技能训练的同时，按照教育部相关精神，增加设计性实验部分，以现实问题作为驱动力来培养学生自主获取和应用新知识的能力，从而培养学生独立思考能力、实验设计能力、实践操作能力和可持续发展能力，满足培养应用型和复合型人才的要求。强调全套教材内容的整体优化，并注重不同教材内容的联系与衔接，避免遗漏和不必要的交叉重复。

三、对接职业资格考试，“教考”“理实”密切融合

本套教材的内容和结构设计紧密对接国家执业中药师职业资格考试大纲要求，实现教学与考试、理论与实践的密切融合，并且在教材编写过程中，吸收具有丰富实践经验的企业人员参与教材的编写，确保教材的内容密切结合应用，更加体现高等教育的实践性和开放性，为学生参加考试和实践工作打下坚实基础。

四、创新教材呈现形式，书网融合，使教与学更便捷更轻松

全套教材为书网融合教材，即纸质教材与数字教材、配套教学资源、题库系统、数字化教学服务有机融合。通过“一书一码”的强关联，为读者提供全免费增值服务。按教材封底的提示激活教材后，读者可通过 PC、手机阅读电子教材和配套课程资源（PPT、微课、视频等），并可在线进行同步练习，实时收到答案反馈和解析。同时，读者也可以直接扫描书中二维码，阅读与教材内容关联的课程资源，从而丰富学习体验，使学习更便捷。教师可通过 PC 在线创建课程，与学生互动，开展在线课程内容定制、布置和批改作业、在线组织考试、讨论与答疑等教学活动，学生通过 PC、手机均可实现在线作业、在线考试，提升学习效率，使教与学更轻松。此外，平台尚有数据分析、教学诊断等功能，可为教学研究与管理提供技术和数据支撑。需要特殊说明的是，有些专业基础课程，例如《药理学》等 9 种教材，起源于西方医学，因篇幅所限，在本次双语教材建设中纸质教材以英语为主，仅将专业词汇对照了中文翻译，同时在中国医药科技出版社数字平台“医药大学堂”上配套了中文电子教材供学生学习参考。

编写出版本套高质量教材，得到了全国知名专家的精心指导和各有关院校领导与编者的大力支持，在此一并表示衷心感谢。希望广大师生在教学中积极使用本套教材和提出宝贵意见，以便修订完善，共同打造精品教材，为促进我国高等中医药院校中药学类专业教育教学改革和人才培养做出积极贡献。

全国高等中医药院校中药学类专业双语规划教材
建设指导委员会

数字化教材编委会

主　编　赵　骏　胡冬华

副主编　韩　波　杨　静　张园园　徐春蕾　盛文兵

编　者　（以姓氏笔画为序）

方玉宇（成都中医药大学）　尹　飞（天津中医药大学）
刘晓芳（山西中医药大学）　杨　静（河南中医药大学）
李贺敏（南京中医药大学）　李嘉鹏（天津中医药大学）
肖新生（湖南科技学院）　张　薇（北京中医药大学）
张园园（北京中医药大学）　张艳春（安徽中医药大学）
林玉萍（云南中医药大学）　赵　骏（天津中医药大学）
胡冬华（长春中医药大学）　施小宁（甘肃中医药大学）
姚惠文（湖北中医药大学）　桂清文（湖南农业大学）
贾鹏昊（天津中医药大学）　徐秀玲（浙江中医药大学）
徐春蕾（南京中医药大学）　郭占京（广西中医药大学）
盛文兵（湖南中医药大学）　韩　波（成都中医药大学）

前 言

有机化学实验与指导是中药学类、药学类及相关专业的一门重要实验基础课程，是后续课程和今后工作学习所必须掌握的基本知识及技能。本书立足于双语教学实际需要，以实验技能训练为主，是一本适合全国高等中医药院校中药学类专业双语教学使用的教材。

本教材由全国15所中医药院校的有机化学专家、教授联合编译，是依据各中医院校相关专业有机化学实验课开设的实际需要和国家教育培养创新型人才要求编写，涵盖了全国高等中医药院校中药学类、药学类及相关专业在教学中比较成熟的实验，适度增加选用实验。

本教材为中英文对照，力求在保持英文实验表述特色的基础上，同时兼顾两种语言的表达特点。全书内容分为四个部分：第一部分有机化学实验的一般知识，以意译为主；第二部分基本操作，第三部分有机化合物合成实验，第四部分天然有机物提取纯化及简单药物合成，基本上采用互译方式。内容编写采用先英文，后中文对照形式，目的是通过本教材的学习，同时使学生专业英语技能得到提升。在操作实验中，中文着重对应实验原理和实验操作，确保学生充分理解实验原理和准确掌握实验操作。

第一章介绍有机化学实验一般知识，包括实验室规则、实验室安全、有机化学实验室常用仪器、常用有机溶剂等；第二章基本操作，把基本操作理论与基本实验技能训练结合在一起，选择8个必须掌握的基本技能操作实验；第三章有机化合物合成实验，选择15个有机化合物合成实验，实验基本为各高校已实施的成熟实验方案，供各院校根据实际情况选择使用；第四章天然有机物提取纯化及简单药物合成，选择多个不同类型的实验。在选编合成和提取实验时，首先考虑有机反应类型的重要性和代表性，其次兼顾一些较新的合成方法和实验技术，并且注意一些基本操作的重复次数，使学生能够熟练掌握基础操作技能。教材后列有附录，主要是常用有机溶剂的物理常数等。

由于水平所限，书中难免存在错误和不妥之处，欢迎读者在使用中提出宝贵意见，以便再版时修订提高。

编 者

2020年4月

Preface

Organic Chemistry Experiment and Guidance is an important experimental basic course for Chinese material medica, pharmacy and other related majors, containing the basic knowledge and skills that must be mastered in subsequent courses and future work. Based on the practical needs of bilingual teaching, this textbook mainly focuses on training experimental skills, which is a bilingual textbook for Chinese medicine major in higher education institutions of Chinese medicine.

This textbook is jointly compiled by organic chemistry experts and professors from 15 universities of Chinese medicine. It is based on the actual needs of organic chemistry experimental courses for universities and colleges of Chinese medicine, covering mature experiments in the teaching process of Chinese materia medica , pharmacy, pharmaceutical engineering and related majors in higher education institutions of Chinese medicine. Some alternative experiments are also included. This book is prepared according to the national requirements of educating and training innovative talents.

This textbook is written in both Chinese and English, and attached great importance to maintain the characteristics of English in experimental description while taking the expression characteristics of both languages into account. The content of the book is divided into four parts: the first part is general knowledge of organic chemistry experiments, which is organized in free-translation way and does not emphasize complete one-to-one correspondence; the second part is basic operations; the third part is organic synthesis experiments; and the fourth part is extraction of natural products and synthesis of simple drugs. They are basically organized by inter-translation. The content is written in English, followed by Chinese. The purpose is to improve students' professional English skills through the study of this textbook. In the experimental parts, Chinese translation focuses on corresponding experimental principles and operations to ensure that students fully understand the experimental principles and accurately grasp the experimental operations.

The first chapter of the book is general knowledge of organic chemistry experiments, including laboratory rules, laboratory safety, common instruments used in organic chemistry laboratories, and standard formats of lab reports, etc. The second chapter is basic operations of organic chemistry experiments, aiming to combine the basic theory with basic experimental skills. Eight basic experimental skills that need to be mastered are selected. The third chapter is organic synthesis experiments, and fifteen organic compounds selected. These experiments are basically mature, which have been conducted in various universities. Each college could make choices according to their actual situations. The fourth chapter is the extraction of natural products and synthesis of simple drugs, choosing nine different types of experiments. When selecting synthesis and extraction experiments, the first consideration should be the importance and representativeness of organic reactions types; some newer synthesis methods and experimental techniques are taken into consideration secondly, and the repetition number of some basic operations is paid attention to, so that students can master the basic operating skills. Appendices are listed at the end of the textbook, including the physical constants of commonly used organic solvents.

Due to our limited level and hasty time, faults and shortages are unavoidable. Readers are welcome to give valuable advice and suggestions so that we can improve during reprinting.

The editor

April 2020

目录 | Contents

Part I　Introduction of Laboratory

第一部分　有机化学实验的一般知识

Part II　Basic Techniques

第二部分　基本操作

Part III Preparative Experiments

第三部分 有机化合物合成实验

Part IV Extraction and Purification of Natural Products and Synthesis of Simple Drugs

第四部分 天然有机物提取纯化及简单药物合成

Part Ⅰ 第一部分

Introduction of Laboratory

有机化学实验的一般知识

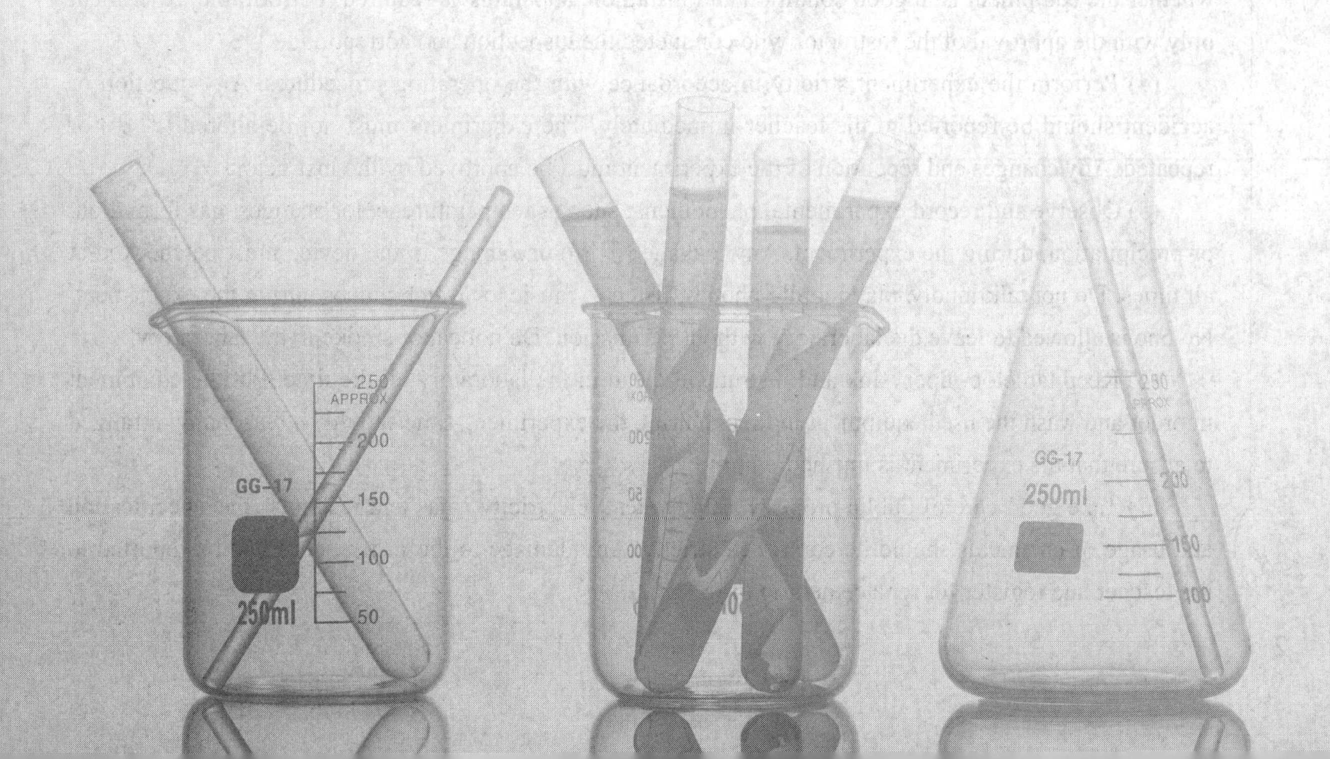

1.1 Safety Rules of Organic Chemistry Laboratory

The inflammable (易燃的), explosive, and volatile (挥发的) characteristics of organic compounds determine that organic experiments are more dangerous than other experiments. Ensuring laboratory safety is a basic requirement in organic chemistry experiments. Thus, students must read the first chapter of this book before entering the laboratory. Relevant contents include the general knowledge of organic experiments and information on toxic and hazardous chemical agents in the appendix. Students must understand laboratory safety regulations, accident prevention, and first-aid measures. When entering the laboratory, students must be familiar with the internal structure of the laboratory, especially the locations of electric switches, fire extinguishers, and emergency exits.

To ensure the normal performance of the experiments and the teaching quality of experimental courses, aside from the above requirements, students must obey the following laboratory rules.

(1) Before the experiment, preview the experiment content carefully; review the relevant theoretical contents in the textbook; clarify the purpose and requirement of the experiment; understand the principle, content and method of the experiment; and write a preview report as required. Understand the experiment notes, possible accidents and preventive (预防的) measures.

(2) Wear proper laboratory clothing, bring the experiment textbook, write a preview report, record and accept the teacher's inspection when entering the laboratory. Familiarize the laboratory environment and follow the safety rules. Dress properly in the laboratory. Slippers, vests and other unsafe or improper clothing are forbidden. Anyone who does not follow the rules is not allowed to carry out the experiment.

(3) Before the experiment, understand the purpose and method of each step, determine the key steps and difficulties in the experiment, and understand the properties of reagent and safety issue. Check whether the equipment is in good condition and install the apparatus as required. Perform the experiment only with the approval of the instructor who conducted the inspection and correction.

(4) Perform the experiment strictly in accordance with the operating procedures. Any question or accident should be reported to the teacher immediately. The experiment must not be altered (改变) or repeated. Any changes and repetition of the experiment must be approved by the instructor.

(5) Observe and record experimental phenomena, such as temperature, color changes, gas formation, or precipitation, during the experiment. Any leakage (泄露) or damage in the device must be checked at all times. Do not talk loudly, play mobile phones, listen to music, or watch videos during the experiment. No one is allowed to leave the laboratory without permission. Do not eat or smoke in the laboratory.

(6) Keep tabletop, floor, sink and instrument clean in the laboratory all the time. Put the equipment in order and wash the used equipment in time. During the experiment, time should be reasonably arranged to ensure that the experiment is finished on time.

(7) Take good care of public property, save water、electricity、gas and reagents. The specification and usage of chemicals should be controlled strictly. Any damage to the instrument must be reported to the teacher and register for replacement in time.

(8) When storing products or recycled liquids in empty bottles, one must develop the habit of timely labeling. The label should indicate the name and time of the substance to avoid the trouble of post-treatment (后处理) or accident caused by the unknown chemical materials. Mixing any chemicals arbitrarily is forbidden to avoid accidents. Waste paper、match stick and liquid waste must not be placed in the sink. After using, please put them into the waste container to avoid blockage or corrosion of the sink and drain. Operations that may produce irritating or toxic gases must be carried out in a fume hood (通风橱).

(9) After the experiment is completed, the experiment records and products should be submitted to the instructor for inspection, registration (登记) and recycling. Please clean the instrument and individual experiment table. Check the water and electricity switch, please inform the instructor to check after confirming safety, and then leave the laboratory with permission.

(10) Students on duty should be arranged for each experiment. Students on duty are responsible for cleaning the public desk、blackboard、floor, checking whether the water and electricity switch is closed, taking out the garbage, closing doors and windows and leaving with the approval of the instructor.

一、实验室规则

有机化合物易燃、易爆和易挥发等特点决定了有机实验比其他实验课更具危险性，保证实验安全是有机化学实验最基本的要求，为此，学生在进入有机实验室之前，必须认真阅读本书第一部分：有机化学实验的一般知识及有关毒性、危害性化学药品的知识。了解实验室安全守则、实验室事故的预防、处理和急救措施等常识；进入实验室后，首先了解实验室的结构，尤其是电闸、灭火器材的位置，熟悉实验室安全出口和紧急逃生路线。

为了保证有机化学实验课能正常、安全、有效地进行，保证实验课的教学质量，除以上要求外，学生还必须遵守下列实验室守则。

1. 实验前要求学生认真预习实验内容，复习理论课教材中有关的内容，明确实验目的和要求，熟悉实验的原理、内容和方法，并按要求写好实验预习报告。了解实验注意事项、可能发生的事故及预防措施。

2. 进入实验室要穿好工作服，带好实验课教材、预习报告和记录本，主动接受指导教师的检查。熟悉实验室环境，遵守实验安全规则，不得穿拖鞋、背心等不安全或不雅观的服装。不遵守规定的不得进行有机化学实验。

3. 实验前要弄清每一步操作的目的、操作方法，实验中的关键步骤及难点，了解所用试剂的性质及应注意的安全问题。检查仪器是否完好无损并按照要求安装实验装置，经指导教师检查、纠正，合格后方可进行下一步的操作。

4. 实验中要严格按操作规程进行，若有疑难问题或意外事故，应立即报请老师解决和处理。不能随意改变和擅自重做实验，如确需改变或重做实验，必须经指导教师同意。

5. 实验过程中，应仔细观察实验现象，如温度、颜色的变化，有无气体、沉淀产生等，并养成及时做记录的良好习惯；随时注意观察装置是否有漏气、破损等现象。不得大声喧哗，不得玩手机、听音乐、看视频等，不得擅自离开实验室，不得在实验室吃东西或吸烟。

6. 实验自始至终要保持桌面、地面、水槽、仪器四净。待用仪器摆放整齐有序，使用过的仪器应及时洗涤。实验过程中，要合理安排好时间确保实验准时结束。

7. 要爱护公物，节约水、电、煤气和药品，严格控制药品的规格和用量。如有仪器或设备的损坏需及时告知指导教师，及时登记更换。

8. 用空瓶盛装产品或回收液时，必须养成及时贴标签的习惯。标签上注明物质名称、时间，以免后续处理麻烦或不知内盛何物而引发事故。杜绝把各种化学药品任意混合，以免发生意外事故。废纸、火柴棒和废液等不得放在水槽内，以防水槽和下水道堵塞或腐蚀，实验后倒入污物桶内。可能产生刺激性或有毒气体的实验操作必须在通风橱内进行。

9. 实验完成后，将实验记录及产品交由指导教师检查、登记并回收，清洗实验仪器，将个人实验台面打扫干净，检查水电开关，确认安全后请指导教师检查，合格后方可离开实验室。

10. 每次实验需要安排值日生。值日生负责实验室的公共台、黑板及地面的清洁，倒垃圾，检查水电开关是否关闭，关门关窗。

1.2 Laboratory Safety

1.2.1 General guides

Many of the drugs used in experiments are flammable, explosive, corrosive or toxic. Accidents such as fire, explosion, poisoning and burns may occur under incorrect operations. In addition, many chemical reactions are carried out under different conditions as high temperature, high pressure, low temperature or low pressure. Various heat sources, electrical appliances and instruments are needed. Improper operation may cause accidents such as electric shock, fire or explosion. Therefore, we should be aware that organic chemistry laboratories are potentially dangerous places. Operators must pay attention to avoid accidents. As long as we learn the properties of the drugs and instruments, and follow the operating rules, we could carry out the experiments safely. In order to avoid accidents or handle them in time, we should follow the guides.

(1) Before experiments, make sure to be familiar with the characteristics, and especially the toxicity of the reagents.

(2) Make sure the equipment work properly. Distillation, reflux and other heating devices are required to connect with air.

(3) During the experiment, pay attention to the instruments and the reactions all the time. Don't leave your work at will.

(4) Do not heat flammable and volatile regents in open containers. Masks and gloves should always be used in potentially dangerous experiments.

(5) All reagents must be kept in safe places and should not be deserted at will. The emitted harmful gases should be handled in the fume hood to avoid environment pollution and to protect personal health.

(6) Whenever glass tube or thermometer is inserted into the plug, make sure that the hole and glass are smooth. Wrapped them with cloth, swear a little glycerin and insert them in slowly while rotating. The hand holding the glass tube should be as close as to the plug to avoid cutting.

(7) Wash hands after experiments. No smoking or eating in the laboratory.

(8) Be familiar with safety appliances like fire extinguishers, sand buckets and first aid kits. They are not allowed to be moved to other places at will.

1.2.2 Prevention and treatment of common laboratory accidents

1.2.2.1 Electrical shock Contacting power supply directly may bring electrical shock. Electrical current will cause tissue damage, dysfunction and even death. The longer time the shock, the more serious damage it is. People may be frightened, palpitated, paled or dizzy under light electricity exposure. While coma, shock, arrhythmia, cardiac or respiratory arrest, coking of the skin and organized necrosis may

occur under severe electric shock. In order to use electrical appliances better and avoid electric shock, follow the rules.

(1) Check electrical equipment regularly and follow operation instructions.

(2) Live work should be avoided whenever possible. If have to, safety measures must be guaranteed.

(3) Static electricity may cause hazards from mild electric shock. To eliminate static electricity, measures must be taken to limit its production or accumulation. It is also possible to use devices fended off the electricity.

(4) Avoid contacting the conductive equipment. No wet hands are allowed to touch electric devices. To prevent electric leakage, the out surface of the devices should link the ground wire. After the experiments, power supply should be switched off and plugs are detached.

In case any electric shock, rescue should be carried out immediately without any delay. Switch off power supply and separate the electric stuff from human body with insulators such as wooden sticks. Whenever heartbeat stop happens, artificial breath like extra cardiac compression is exerted until heartbeat resumes or the patient is sent to the hospital.

1.2.2.2 Ignition Generally, the chemical reagents with flash point below 25℃ are recognized as flammable chemical reagents. Most of them are volatile and burn when exposed to open fire. Most of the organic solvents are flammable. Ignition is a common accident in organic experiments. It is necessary to keep the fire source and solvent away and to avoid direct heating with open fire as far as possible.

Pay attention to the following matters.

(1) Large quantity of flammable solvents should be stored in safe cabinets instead of kept in the laboratory or near the fire.

(2) Do not store, heat or evaporate flammable solvents in open containers as beakers. Keep them away from fire sources.

(3) Do not heat flasks with flame directly. Select the proper heat source when heating flasks.

(4) Avoid the escape of flammable solvents into air. In case escape happens, extinguish the fire source and release the organic vapor immediately.

(5) Do not dispose flammable or volatile liquid into waste tanks. Keep away from fire source whenever disposing. It would be preferred to perform pouring in the fume hood. Recycle the organic solvents when the large amount is used.

(6) When distilling or refluxing, zeolites are needed to prevent explosive boiling. Zeolites should not directly add to the superheated solution, otherwise the liquid may boil suddenly, rush out and catch fire. When oil bath is heated, water droplets must be kept away from the hot oil.

(7) When distilling or refluxing, cooling water must be running smoothly. If not, a large amount of hot gas may escape and catch fire.

(8) Extinguish fire temporally or away from fire source when transferring flammable solvents. Do not slant one alcohol lamp to another to ignite it. Alcohol lamp should be covered after use. Do not use alcohol lamps with damaged necks. Turn off the fire, heat source and tap water before leaving the laboratory.

1.2.2.3 Injuries

(1) Chemical reagent burns Corrosive and flammable reagents should be used in strict accordance with the rules. Students need to wear rubber gloves and protective glasses if necessary. Burns should be

treated properly.

Acid-injury: use a large amount of water, 3%~5% sodium bicarbonate solution for skin-wash and 1% sodium bicarbonate solution for eye-wash. If skin stained with concentrated sulfuric acid, wipe the acid with a dry cloth and treat the skin as above.

Alkali-injury: rinse with plenty of water. Washed with 2% acetic acid aqueous and water.

Bromine-injury: wash with water and alcohol, or with 20 g/L sodium thiosulfate aqueous until the skin becomes white. Finally, paste glycerin or scald ointment on the skin.

(2) Cuts improper use of glass instruments may cause cuts. Do not install and separate glass instruments forcefully. When glass cut happens, remove the glass from body firstly. The contaminated blood should be squeezed out. After washing with water, the wound should be covered with iodine liquor and bandaged. If the wound is serious, the main blood vessel should be tightened to prevent massive bleeding, and hospital treatment is required.

(3) Scald heating is frequently used in experiments, therefore scald happens commonly. For light scald, smear ointment to keep the skin safe.

1.2.3 The use of laboratory safety equipment

In order to ensure the safety of organic chemistry laboratory, fire-proof (防火), explosion-proof (防爆), and fire extinction (灭火) are the main approaches. Fire sand buckets (消防沙桶), fire blankets (消防毯) and fire extinguishers (灭火器) are safety equipment commonly used in laboratories. Being familiar with the performance and scope of these safety devices and mastering the method of application and operating correctly can avoid fire accidents and ensure laboratory safety.

1.2.3.1 Fire sand bucket Fire sand bucket is used for fire extinction by covering the object of ignition to isolate fire from air. Fire sand buckets are low in cost and with available materials, which are suitable for initial chemical fires (化学品着火) and Class D related metal fire (D类金属火灾), including sodium metal fire, potassium metal fire, aluminum metal fire, magnesium metal fire, aluminum-magnesium alloys metal fire, and titanium metal fire, etc.

Generally $20m^2$ house should be equipped with about 50 liters of sand. Fire sand also absorb flammable liquids and must be kept dry. When using it, cover the fire sand to isolate from air until the fire is out.

1.2.3.2 Fire blanket Fire blanket has excellent fire extinguishing property and widely application. Place the fire blanket in a conspicuous place which can be quickly taken. To use a fire blanket, remove from its packing, grip the unfolded blanket in front of you with your hands and body protected by it and drape the blanket over the fire. Cover the flame with the fire blanket while turning off the power. After the object of ignition is extinguished and the fire blanket is cooled, wrapped the blanket into a ball and treated as non-burnable garbage. If the person is on fire, shake out the fire blanket and completely wrap it around the burning man.

1.2.3.3 Fire extinguisher Common fire extinguishers include dry powder extinguisher (干粉灭火器), carbon dioxide extinguisher (CO_2灭火器), foam extinguisher (泡沫灭火器) and 1211 fire extinguisher (1211灭火器). It is necessary to be familiar with the characteristics, scope of application, precaution and proper operation of these fire extinguishers for fire prevention and fire extinguishing.

(1) Dry powder fire extinguisher Dry powder fire extinguishers are characterized by ease for use and valid for long. Among them, the portable dry powder fire extinguisher (手提式干粉灭火器)

is the most commonly used in the laboratory. The dry powder fire extinguisher generally divides into two categories: sodium bicarbonate powder (碳酸氢钠干粉) and ammonium phosphate powder (磷酸铵盐干粉). It is suitable for smothering the fire caused by flammable, combustible solid, liquid and gas, and electrical equipment, but it cannot smother metal burning fires. Before using the dry powder fire extinguisher, the bottle should be inverted several times to loosen the powder in the cylinder. Then remove the seal and the insurance pin (保险销), hold the nozzle in left hand and press the handle in right hand. Stand two meters away from the fire flame, push it with right hand and swing the nozzle with left hand, spray the dry powder to cover the burning area until the fire is completely extinguished and avoid the possibility of re-ignition. If the burning liquid flowing around, aim at the root of the flame and spray the dry powder around until the flame completely extinguished.

(2) Foam fire extinguisher　Foam fire extinguishers are suitable for smothering fires caused by oil fires (油类火灾) and solid combustibles (固体可燃物) such as wood, fiber and rubber. The foam fire extinguisher is internally filled with sodium bicarbonate (碳酸氢钠溶液) and aluminum sulfate solution (硫酸铝溶液) which containing foaming agents. When in use, invert the barrel to make the two solutions react to form sodium hydrogen sulfate (硫酸氢钠), aluminum hydroxide (氢氧化铝) and a large amount of carbon dioxide. Thus the pressure inside the fire extinguisher cylinder suddenly increased, and a large amount of carbon dioxide foam was ejected. Usually foam fire extinguishers are not used unless big fire, because it sprayed a large amount of pollutants such as sodium sulfate and aluminum hydroxide, caused trouble for post-processing. When using the foam fire extinguisher, it should be noted that people should stand up the wind and try to be close to fire source, because its spray distance is only 2~3 meters. Spray from the most dangerous side of the fire and then gradually move to the other side, not to leave sparks. Besides, hold the wooden handle of the nozzle to avoid frostbite. Too much carbon dioxide in the air is harmful to human body, so it should be ventilated immediately.

(3) Carbon dioxide fire extinguisher　Carbon dioxide fire extinguisher has the advantages of high extinguishing performance, low toxicity, small corrosion, no trace after extinguishing, which are commonly used in laboratory. It is suitable for all kinds of flammable liquid and combustible gas fire, and smothering the fire of valuable equipment, instruments, books and files, and low-voltage electrical equipment. There are two types of carbon dioxide extinguishers: switch type (开关式) and guillotine type (闸刀式). The steel cylinder contain compressed liquid carbon dioxide, when in use, pull out the insurance pin first, and then lift the fire extinguisher in one hand, hold on the handle of carbon dioxide horn in the other hand. Please don’t hold the horn! Avoid frostbite. Be noticed: once the guillotine fire extinguisher is opened, it can’t be turned off again. Make ready before use.

(4) 1211 fire extinguisher　1211 fire extinguisher has the advantages of high fire-fighting efficiency, low toxicity, small corrosiveness, long storage and no deterioration, no trace of fire-fighting, no pollution of protected materials, and good insulation performance. 1211 fire extinguisher is mainly used to extinguish fire caused by flammable liquid, gas, metal, and smother initial fire of precision instruments, valuable goods, precious cultural relics, books and archives. It is a kind of halon fire extinguisher (卤代烷灭火器), which is produced and widely used in our country. When in use, first remove the safety pin, and then hold the pressure handle tight to spray. It should be noticed that when extinguishing the fire, the extinguisher should be put out in an upright position, not horizontally or upside down. The nozzle should aim at the root of the flame, from far to near, and push forward the fire in a rapid and

flat manner.

Whatever kinds of extinguisher you use, extinguish fire from periphery to the center. It is important to note that oil baths (油浴) and organic solvents should never be covered with water when they are on fire, as this will only spread the flames.

二、实验室安全

（一）实验室一般安全事项

在有机化学实验中，许多药品是易燃、易爆、有腐蚀性或有毒的危险品，因此稍有不慎就有可能发生火灾、爆炸、中毒、烧伤等事故。另外，在化学反应时，常需要在高温、高压、低温、低压等不同的条件下进行，需要使用各种热源、电器及仪器，若操作不小心，就可产生触电、火灾、爆炸等事故。因此，必须充分认识到有机化学实验室是潜在危险的场所。实验者必须树立安全第一的观念，重视预防就可以避免发生事故。只要认真预习和熟悉实验中使用的药品和仪器的性能、用途、可能出现的问题及预防措施，加强安全意识和安全措施，并严格执行操作规程，就能有效地维护人身和实验室的安全，确保实验的顺利进行。为了防止事故的发生，以及发生事故后及时处理，我们应高度重视以下事项，并切实执行。

（1）实验前认真预习，了解实验所用药品的性能及其危害。

（2）实验开始前应检查仪器是否完整无损，装置是否正确稳妥。注意蒸馏、回流等装置以及加热用仪器，一定要和大气接通。

（3）实验过程中应该经常注意仪器有无漏气、破裂，反应进行是否正常，不得随意离开岗位。

（4）易燃、易挥发药品，避免放在敞口容器中加热；有可能发生危险的实验，在操作时应使用防护眼镜、面罩和手套等防护设备。

（5）实验中所有药品，不得随意散失和遗弃。对反应中产生有害气体的实验，应在通风柜中处理或按规定处理，以免污染环境，影响身体健康。

（6）玻璃管或温度计插入塞中时，应先检查塞孔大小是否合适，玻璃切口是否光滑，用布裹住并涂少许甘油等润滑剂后再缓缓旋转而入。握玻璃管的手应尽量靠近塞子，以防因玻璃管折断而割伤。

（7）实验结束后要及时洗手；严禁在实验室内吸烟或饮食。

（8）要熟悉安全用具如灭火器、沙桶以及急救箱的放置地点和使用方法，并妥善保管。安全用具及急救药品不准移作他用，或随意挪动存放位置。

（二）常见实验室事故预防和处理

1. 触电　触电是由于人体直接接触电源产生的，人体受到一定量的电流会导致组织损伤和功能障碍甚至死亡。触电时间越长，人体所受的电损伤越严重。轻者惊吓、心悸、面色苍白、头晕乏力。重者立即出现昏迷、休克、心律失常或心脏骤停、呼吸停止并伴着电击部位皮肤的电灼伤、焦化，并有组织坏死。

为了更好地使用电器和电能，防止触电事故的发生，必须采取一些安全措施。

（1）经常定期检查各种电器设备，如发现故障或不符合有关规定的，应及时处理。严格遵守各种电气设备操作使用制度和说明。

（2）尽量不要带电工作，特别是在危险场所，禁止带电工作。如果必须带电工作，应采取必要的安全措施。

（3）静电可能引起危害，轻则可使人受到电击，重则引起爆炸与火灾，引起严重后果。消除静电首先应尽量限制静电电荷的产生或积聚。也可采用性能可靠的漏电保护器。

（4）使用电器时，应防止人体与电器导电部分直接接触，不能用湿的手或手握湿物接触电插头。为了防止触电，装置和设备的金属外壳等都应连接地线。实验完后先切断电源，再将连接电源的插头拔下。

万一发生触电事故，要立即展开施救，迅速切断电源，拉开电闸或用木棍等不导电物将电源与人体分开。立即行人工呼吸，心跳停止时，应立即施行心外按压，并坚持不懈，至复苏或送医院救治。

2. 着火　通常将闪点在25℃以下的化学试剂列为易燃化学试剂，易燃试剂多数是易挥发的液体，遇到明火即可燃烧。有机实验使用的有机溶剂大多数是易燃试剂，着火是有机实验中常见的事故。因此，必须注意防火，防火的基本原则是使火源与溶剂尽可能远离，并尽量避免用明火直接加热。

实验中要注意以下几点。

（1）数量较多的易燃有机溶剂应存放在危险药品橱内，而不能放在实验室内，更不能放置在灯火附近。

（2）切勿用烧杯等敞口容器存放、加热或蒸发易燃有机溶剂，应该远离火源。

（3）避免用火焰直接加热烧瓶，加热时，要根据实验需求及易燃有机溶剂的特点选择理想的热源。

（4）要尽量避免易燃溶剂的气体外逸，若有外逸时要及时灭掉火源，立即排出室内的有机蒸气。

（5）易燃及易挥发物，不得倒入废物缸内，倾倒易燃液体时应远离火源，最好在通风橱中进行，量大时要专门回收。

（6）蒸馏或回流液体时切记放入沸石，以防溶液因过热暴沸而冲出；切记不能在过热溶液中补加沸石否则会导致液体突然沸腾，冲出瓶外而引起火灾事故。油浴加热时，应绝对避免水滴溅入热油中。

（7）蒸馏或回流时，冷凝水要保持畅通，若冷凝管忘记通水，大量蒸气来不及冷凝而逸出遇到火源，也易造成火灾。

（8）在反应中添加或转移易燃有机溶剂时，应注意暂时熄火或远离火源。切忌斜持一只酒精灯到另一只酒精灯上去点火。酒精灯用毕应立即盖灭，避免使用灯颈已经破损的酒精灯。离开实验室时，一定要关闭火源、热源和自来水。

3. 受伤

（1）化学试剂灼伤　一些化学试剂有腐蚀性或容易灼伤皮肤，使用时应严格按照操作规程，必要时可戴上橡胶手套和防护眼镜。如意外发生灼伤时，应视具体情况进行处理。

酸灼伤：立即用大量水冲洗，然后用3%~5%碳酸氢钠溶液处理皮肤。若溅入眼内，用大量水冲洗后，用1%碳酸氢钠溶液处理。浓硫酸沾上皮肤要立即用干布擦去，然后用上述方法处理。

碱灼伤：立即用大量水冲洗，然后用2%醋酸溶液洗，最后用水洗。

溴灼伤：溴灼伤皮肤时，应用水冲洗再用酒精擦洗或用20g/L硫代硫酸钠溶液洗至灼伤处呈白色，然后涂上甘油或烫伤油膏。

（2）割伤　玻璃仪器使用不当造成破损时，可能会引起割伤。因此在玻璃仪器的安装、拆卸中应小心操作，不能使用蛮力。如出现玻璃割伤，应先把伤口处的玻璃碎片取出，挤出污血，用蒸馏水洗后涂上碘酒，用绷带扎住；大伤口应先按紧主血管，以防止大量出血；伤势严重者赶快

送医院处理。

（3）烫伤 实验中常用加热，容易发生烫伤事件，特别是蒸汽的烫伤。轻伤皮肤未破时，可涂以烫伤油膏。

（三）实验室安全器材及使用

为了保证有机化学实验室的使用安全，防火防爆和灭火是最主要的措施。消防沙桶、消防毯和灭火器等是实验室常用的安全器材。熟悉消防沙桶、防火毯、灭火器的性能及适用范围，掌握其使用方法并能正确地进行相关操作可避免火灾等事故的发生，确保实验室安全。

1. 消防沙桶 消防沙是通过覆盖以隔绝着火物与空气接触而达到灭火的目的。消防沙桶费用低，材料易得，适用于火势初起的化学品着火和D类金属火灾，如钠、钾、铝、镁、铝镁合金、钛等金属引起的火灾等。

一般来说 $20m^2$ 的房子应该配备50L左右的沙子。消防沙还有吸纳易燃液体的功能，因此要保持其干燥。使用时将消防沙覆盖着火物品隔绝空气直到火被熄灭为止。

2. 消防毯 消防毯具有优良灭火的性能，适用范围广，要将消防毯放置于比较显眼且能快速拿取的地方。当发生火灾时，快速取出灭火毯，将灭火毯轻轻抖开，作盾牌状拿在手中。将灭火毯覆盖在火焰上，同时切断电源或气源，并采取积极灭火措施直至着火物体完全熄灭。待着火物体熄灭，灭火毯冷却后，将毯子裹成一团，作为不可燃垃圾处理。如果人身上着火，将毯子抖开，完全包裹于着火人身上灭火。

3. 灭火器 常见的灭火器有干粉灭火器、二氧化碳灭火器、泡沫式灭火器以及1211灭火器等四类。熟悉这几种常用的灭火器的特点、适用范围、注意事项以及正确操作和使用灭火器进行防火、灭火是很有必要的。

（1）干粉灭火器 干粉灭火器特点是使用方便、有效期长。其中手提式干粉灭火器是目前实验室最常用的灭火器。干粉灭火剂一般分为碳酸氢钠干粉和磷酸铵盐干粉两大类。其适用于扑救各种易燃、可燃固体、液体和气体火灾以及电器设备的火灾，但不能扑救金属燃烧火灾。使用干粉灭火器前应将瓶体颠倒几次，使筒内干粉松动，然后除掉铅封，拔掉保险销，左手握着喷管，右手提着压把，在距火焰2m的地方，右手用力压下压把，左手拿着喷管左右摇摆，喷射干粉覆盖燃烧区，直至把火全部扑灭并避免复燃。如果被扑救的液体火灾呈流淌式燃烧时，应对准火焰根部由近而远，并左右扫射，直至把火焰全部扑灭。

（2）泡沫式灭火器 泡沫式灭火器适用于扑救各种油类火灾和木材、纤维、橡胶等固体可燃物火灾。泡沫式灭火器内部分别装有含发泡剂的碳酸氢钠溶液和硫酸铝溶液，使用时将筒身颠倒，两种溶液即反应生成硫酸氢钠、氢氧化铝及大量二氧化碳。灭火器筒内压力突然增大，大量二氧化碳泡沫喷出。非大火通常不用泡沫式灭火器，因为泡沫式灭火器喷出大量的硫酸钠，氢氧化铝污染比较严重，给后续处理带来麻烦。使用泡沫式灭火器时应该注意，人要站在上风处，尽量靠近火源，因为其喷射距离只有2~3m，要从火势蔓延最危险的一边喷起，然后逐渐移动，不要留下火星。手要握住喷嘴木柄，以免被冻伤。如果在空气中的二氧化碳含量过多，会对人体产生不利影响，所以在空气不畅通的场合，喷射后应立即通风。

（3）二氧化碳灭火器 二氧化碳灭火器的灭火性能高、毒性低、腐蚀性小、灭火后不留痕迹，是实验室比较常用的灭火器。其适用于各种易燃、可燃液体和可燃气体火灾，还可扑救贵重设备、仪器仪表、图书档案和低压电器设备等初起火灾。二氧化碳灭火器有开关式和闸刀式两种。它的钢筒内装有压缩的液态二氧化碳，使用时先拔去保险销，然后一手提灭火器，一手应握在喷二氧化碳喇叭筒的把手上，不能手握喇叭筒，避免冻伤。需要注意的是：闸刀式灭火器一旦打开后，就再也不能关闭了。因此，在使用前要做好准备。

（4）1211灭火器　1211灭火器具有灭火效率高、毒性低、腐蚀性小、久储不变质、灭火后不留痕迹、不污染被保护物、绝缘性能好等优点。1211灭火器主要适用于扑救易燃、可燃液体、气体、金属及带电设备引起的火灾；扑救精密仪器、仪表、贵重的物资、珍贵文物、图书档案等初起火灾。其是利用装在筒内的氮气压力将1211灭火剂喷射出灭火，是我国目前生产和使用最广的一种卤代烷灭火剂，以液态罐装在钢瓶内。使用时，首先拔掉安全销，然后握紧压把进行喷射。但应注意，灭火时要保持直立位置，不可水平或颠倒使用，喷嘴应对准火焰根部，由远及近，快速向前平推进扫射。

无论用何种灭火器，皆应从火的四周开始向中心扑灭。需要注意的是油浴和有机溶剂着火时绝对不能用水浇，因为这样反而会使火焰蔓延开来。

1.3 Common Instruments in Organic Chemistry Laboratory

1.3.1 Glassware, clean and dry

1.3.1.1 Common glassware The glassware (玻璃仪器) is generally divided into two types: common glassware and standard glass-grinding glassware. The common glassware commonly used in the lab includes a conical flask, a beaker, a Büchner funnel, a Büchner flask, a common funnel, a separating funnel, and the like, as shown in Figure 1-1 (a). The common standard mill instrument includes a round bottom flask, a three-necked flask, a distillation head, a condenser, and a receiving tube, and the like, as shown in Figure 1-1 (b). The uses of glassware are shown in Table 1-1.

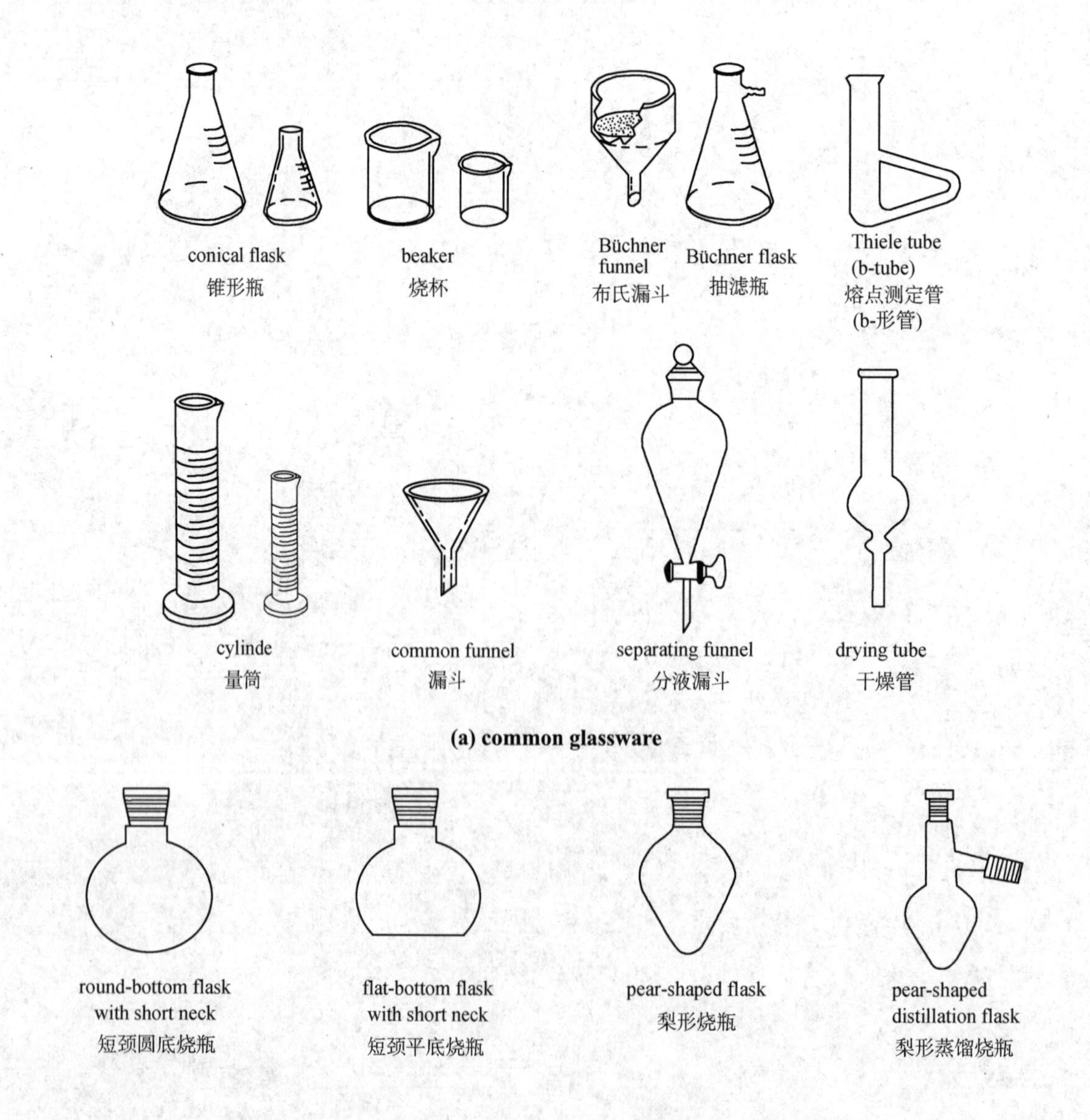

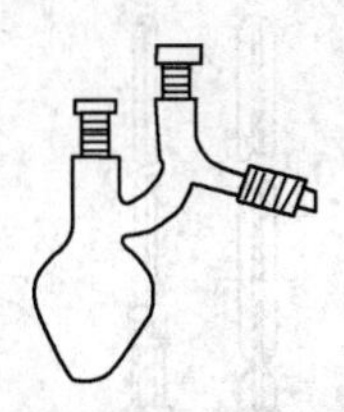
pear-shaped claisen flask
梨形克氏蒸馏瓶

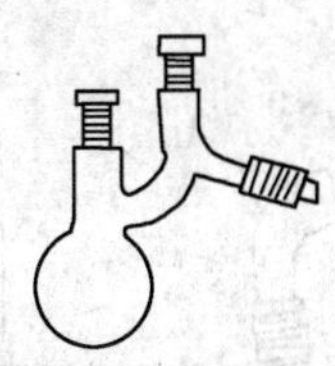
circular claisen flask
圆形克氏蒸馏瓶

circular distillation flask
圆形蒸馏烧瓶

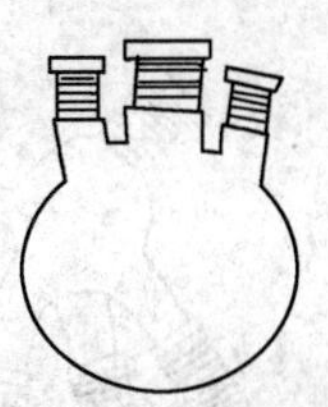
straight three-mouthed flask
直形三口烧瓶

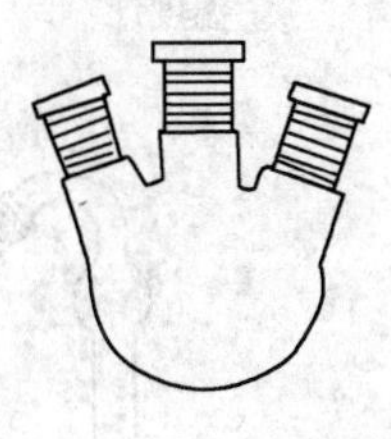
oblique three-mouthed flask
斜形三口烧瓶

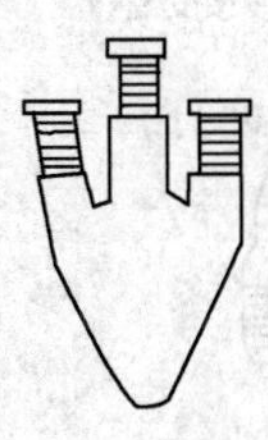
pear-shaped three-mouthed flask
梨形三口烧瓶

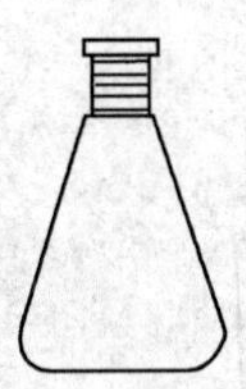
triangular flask
三角烧瓶

aspirator
抽滤管

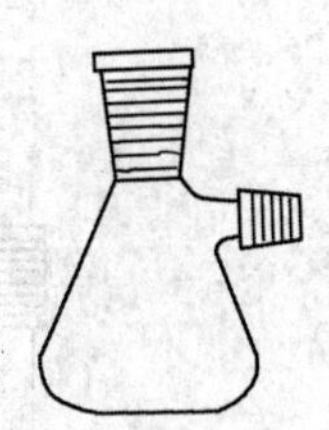
Büchner flask
抽滤瓶

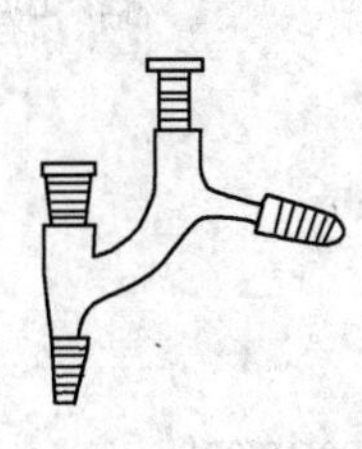
Kirschner distillation head 75°
克氏蒸馏头 75°

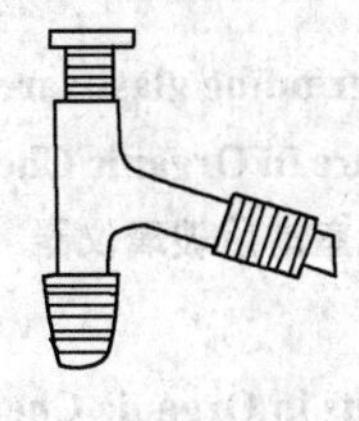
distillation head 75°
蒸馏头 75°

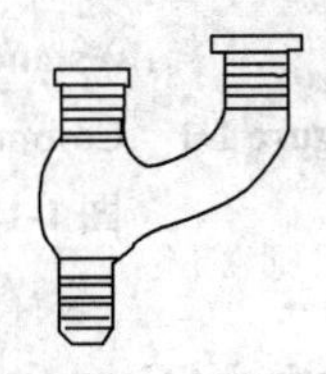
two-port connecting pipe
二口连接管

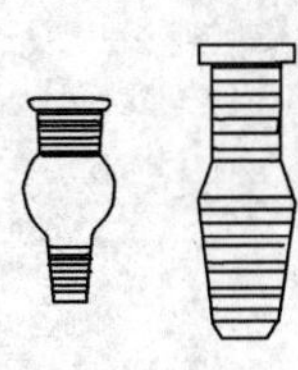
joint
接头

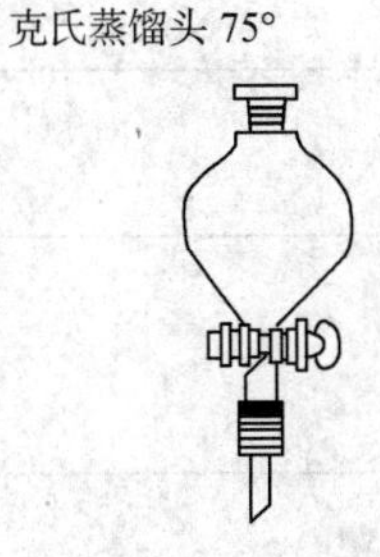
separating funnel
分液漏斗

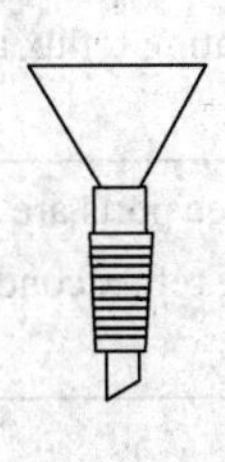
common funnel
漏斗

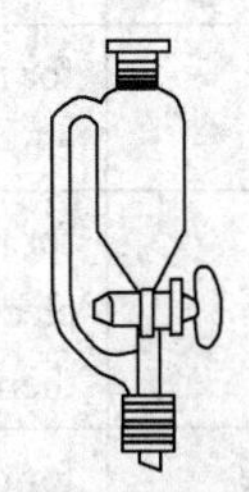
constant pressure dropping funnel
恒压滴液漏斗

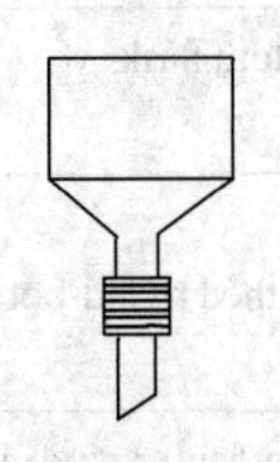
sand core funnel
砂芯漏斗

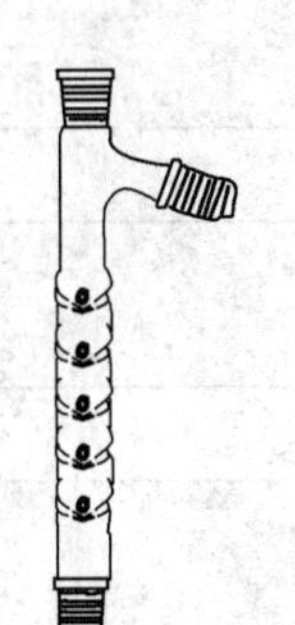
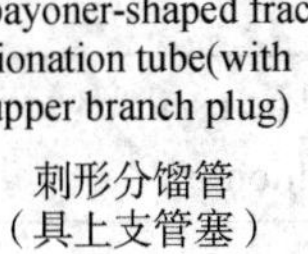
bayoner-shaped fractionation tube(with upper branch plug)
刺形分馏管（具上支管塞）

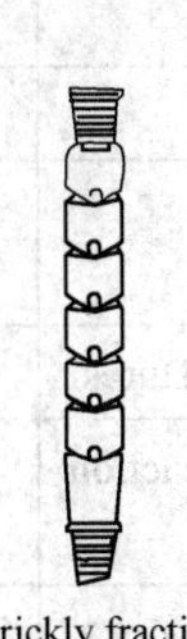
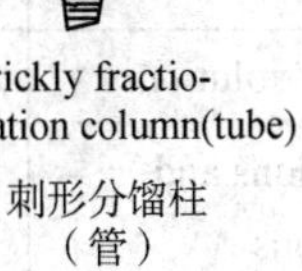
prickly fractionation column(tube)
刺形分馏柱（管）

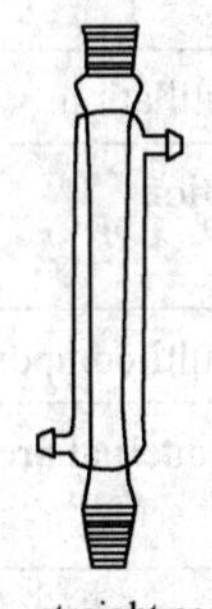
straight condenser pipe
直形冷凝管

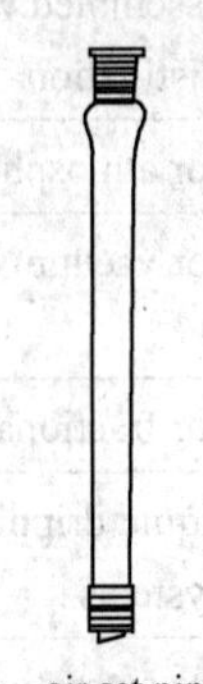
air set pipe
空气冷凝管

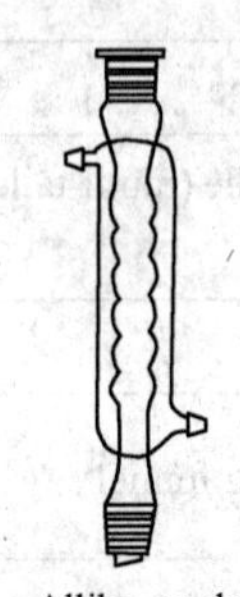
Allihn condenser
球形冷凝管

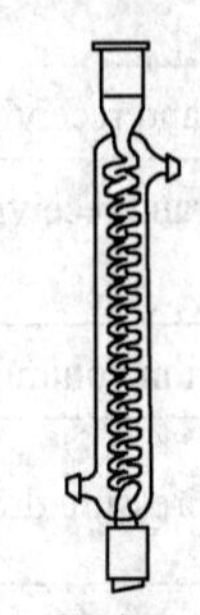
Graham condenser
蛇形冷凝管

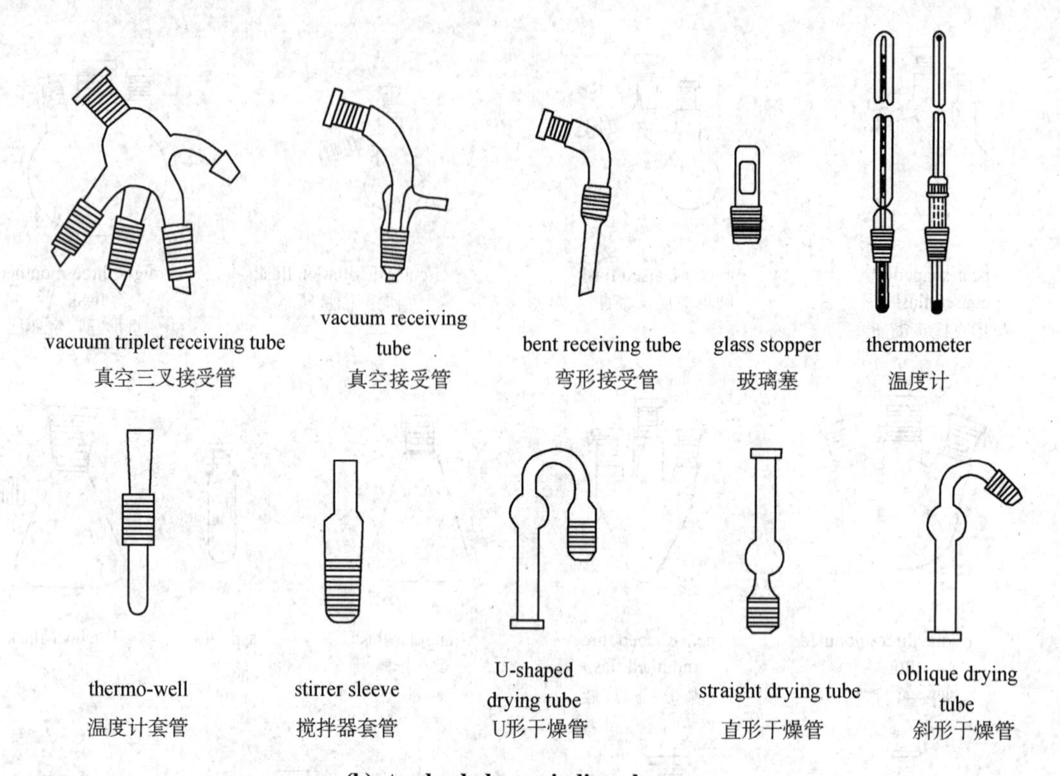

(b) standard glass-grinding glassware

Figure 1-1 Common Glassware in Organic Chemistry Lab

图 1-1 有机实验室常用玻璃仪器

Table 1-1 Application of Common Instruments in Organic Chemistry Experiment

表 1-1 有机化学实验常用仪器的应用范围

Instrument	Applied range	Notes
round-bottom flask	for reaction, heating reflux and distillation, etc	
three-mouthed round-bottom flask	for reaction, three ports are equipped with electric agitator, reflux condenser tube and thermometer, etc	
straight condenser pipe, air set pipe	for distillation	
Allihn (or Graham) condenser	for heating reflux	
distillation head	assembled with round-bottom flask for distillation	
single-strand receiving tube	for atmospheric distillation	
double strand receiving tube (multi-tailed tubing)	for vacuum distillation	
fractionation column	for fractionating multi-component mixtures	
constant pressure dropping funnel	liquid dripping for internal pressure reaction systems	
separating funnel	for extraction and separation of solutions	
conical flask	for storing liquid, mixing solutions and heating small amount of solutions	cannot be used for vacuum distillation

(continued)

Instrument	Applied range	Notes
beaker	for heating, concentrating, mixing and transferring solutions	
cylinder	for measuring liquid	cannot be used for heating
Büchner flask	for decompression filtration	cannot be used for heating
Büchner funnel	for decompression filtration	porcelain
Hirsch funnel	for decompression filtration	porcelain plate for movable circular orifice plate
Thiele tube	for measuring melting point	contain paraffin oil, silicone oil or concentrated sulfuric acid, etc
drying tube	loading of desiccants for anhydrous reaction units	

The size of the standard ground (磨口) instruments' caliber (口径) is usually described by a digital number, which refers to the millimeter integer of the maximum end diameter of the grinder. It is divided into 10, 14, 19, 24, 29, 34, 40, 50models according to the grinding caliber. The child mouth of the same number can be connected to the mother port. When connecting the mother port with a different number of supports, a size joint (转接头) can be added in the middle. Sometimes two sets of numbers are used to indicate the size and length of a caliber, such as 14/30, indicating that the diameter of the grinding mouth is 14mm and the length of the grinding mouth is 30mm. The constant instrument used by the students is generally No. 19 or 24 grinding instruments, and the No. 14 grinding instrument is used in the semi-micro experiment. The No. 10 grinding instrument is used in the micro-experiment. The following points should be noted in using glass instruments.

(1) When you use them, you should handle with care.

(2) Glass instruments should not be directly heated with open fire, cotton cushions should be used when heating.

(3) High temperature cannot be used for heat intolerant glass instruments, such as filter bottles, ordinary funnels, measuring tubes and so on.

(4) After the glass instruments are used, they should be cleaned in time. Especially the standard grinding instruments are easy to bond together when they are placed for a long time without cleaning. If it happens, hot water can be used to ironing the bonding place or blowing the mother grinding mouth with hot air to make it expand and fall off. Moreover, a gavel can also be used to gently knock on the bonding place to make it fall off. Air-dry is best for a cleaned glass instrument.

(5) After cleaning cocks or plugs, paper should be clamped or smeared at the joint between the plug and the grinder to prevent bonding.

(6) The grinding place of the standard grinding instrument should be clean without solid materials stuck. When cleaning, you should avoid scrubbing the grinding mouth with decontamination powder. Otherwise, the grinding connection will not be tight, or even damaged.

(7) When installing the instrument, it should be horizontal or vertical. The grinding joint should not be subjected to skew stress to avoid being broken.

(8) When using glass instruments, the grinder generally does not need to be coated with lubricant

to avoid adhesion to reactants or products. However, when using strong bases in the reaction, lubricant should be used to avoid adhesion caused by alkaline corrosion. When conducting vacuum distillation, lubricant should also be applied to the grinding connection to ensure the device with good leak proofness.

(9) When using thermometers, special care should be taken to avoid washing the hot thermometer with cold water. The thermometer, especially the mercury ball, should be cooled to near room temperature before washing. Remember that the thermometer must not be used as a glass rod to stir to avoid damaging the thermometer.

1.3.1.2 Lab glassware cleaning basics It's generally easier to clean up glassware if you do it right away. When a detergent is used, it's usually one designed for lab glassware, such as Liquinox or Alconox. Usually, neither detergent nor tap water are required. You can rinse the glassware with proper solvents. Then finish up with a couple of rinses with distilled water, followed by final rinses with deionized water.

(1) General chemicals washing ① Water soluble solutions (e. g., sodium chloride or sucrose solutions): Rinse three to four times with deionized water, then put the glassware away. ② Water insoluble solutions (e. g., solutions in hexane or chloroform): Rinse two to three times with ethanol or acetone. Rinse three to four times with deionized water, then put the glassware away. In some situations, other solvents need to be used for the initial rinse. ③ Strong acids (e. g., concentrated HCl or H_2SO_4): In the fume hood, carefully rinse the glassware with copious volumes of tap water. Rinse three to four times with deionized water, then put the glassware away. ④ Strong bases (e. g., 6mol/L NaOH or concentrated NH_4OH): In the fume hood, carefully rinse the glassware with copious volumes of tap water. Rinse three to four times with deionized water, then put the glassware away. ⑤ Weak acids (e. g., acetic acid solutions or 0.1mol/L/1mol/L HCl or H_2SO_4): Rinse three to four times with deionized water before putting the glassware away. ⑥ Weak bases (e. g., 0.1~1mol/L NaOH or NH_4OH): Rinse thoroughly with tap water to remove the base, then rinse three to four times with deionized water before putting the glassware away.

(2) Special glassware washing Glassware used for organic chemistry rinse the glassware with appropriate solvents. Use deionized water for water-soluble contents. Use ethanol for removing ethanol-soluble contents, followed by deionized water. Rinse with other solvents as needed, followed by ethanol and deionized water. If the glassware requires scrubbing, scrub with a brush using hot soapy water, rinse thoroughly with tap water, followed by deionized water. ① Burettes: Wash with hot soapy water and rinse thoroughly with tap water, then rinse three to four times with deionized water. Be sure the final rinses flow out of the glass. Burettes need to be thoroughly cleaned up before used for quantitative analysis in the laboratory. ② Pipets and volumetric flasks: In some cases, you may need to soak the glassware over night in soapy water. Clean up pipets and volumetric flasks with warm soapy water. The glassware may require scrubbing with a brush. Rinse with tap water followed by three to four rinses with deionized water.

(3) Drying or not drying ① Not drying glassware: If the concentration of the final solution is affected by the water, it is necessary to rinse the glassware with the solution for three times. ② Drying glassware: It is inadvisable to dry glassware with a paper towel or forced air since this can introduce fibers or impurities that can contaminate the solution. Normally, you can allow glassware to air dry on the shelf. Otherwise, if water is filled to the glassware, it is fine to leave it wet (unless it will affect the concentration of the final solution). If glassware is to be used immediately after washing and must be dry, rinse it two to three times with acetone. This will remove any water since water will evaporate quickly. While it's not a good idea to blow air into glassware to dry it, sometimes you can apply a vacuum to evaporate the solvent.

(4) Additional tips Remove stoppers and stopcocks when they are not in use. Otherwise, they may "freeze" in place. You can de-grease ground glass joints by wiping them with a lint-free towel soaked with ether or acetone. Wear gloves and avoid breathing the fumes. The deionized water rinse should form a smooth sheet when poured through clean glassware. If this is not seen, more aggressive cleaning methods may be needed.

1.3.2 Common equipments and supporting instruments and their use

There are a lot of electrical equipment in the laboratory. Please pay attention to safety and keep the equipment clean. Do not spill medicine on the equipment.

1.3.2.1 Electro-thermostatic blast oven The laboratory generally uses an electro-thermostatic blast oven, which is mainly used for drying glass instruments or non-corrosive, thermally stable drugs, as shown in Figure 1-2.

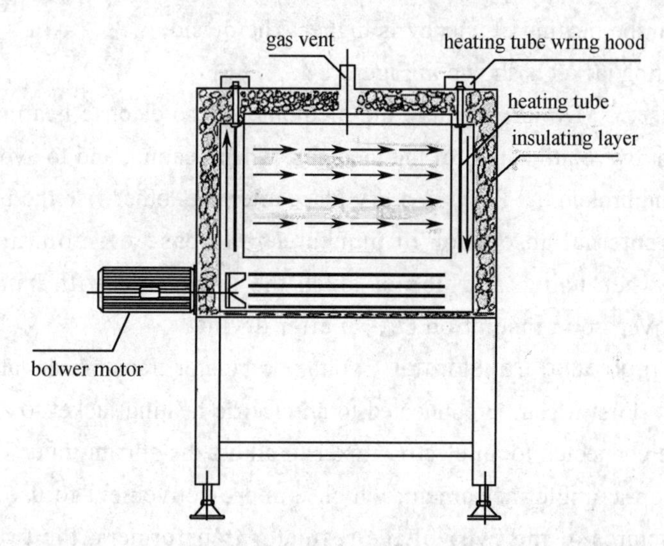

Figure 1-2 Electric Drying Oven With Forced Convection (对流)

图 1-2 强制对流电烘箱

Precautions for use: ① The drying cabinet will be placed at level ground of drying room without active gas. ② The temperature should be adjusted before using (the glass baking instrument is generally controlled at 100~110℃). ③ The freshly washed instrument should drain off the water and put it into the oven without dripping. ④ After removing the hot instrument, do not touch the cold object immediately to prevent the instrument from cracking. ⑤ Instruments with cocks or plugs should be removed and then put into the oven to dry.

1.3.2.2 Glass instrument steam air dryer for Lab

It is a device for rapid drying of glass instruments, such as used to dry test tubes and beakers, etc. The outside drawing of glass instrument steam air dryer for LAB is shown in Figure 1-3.

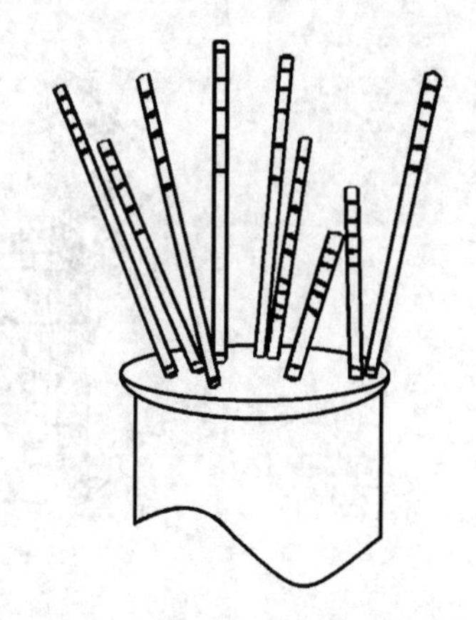

Figure 1-3 Glass Instrument Steam Air Dryer for Lab

图 1-3 实验室用玻璃仪器空气干燥器

Precautions for use: In the use of the dryer, the glass should

be cleaned, drained, and then inserted into the dryer's porous metal tube. The dryer blows the cold air first, then the hot air, and finally the cold air to cool the glass instrument without water vapor condensing to room temperature. The air dryer should not be heated for a long time to avoid burning out the motor and electric wire.

1.3.2.3 Electric heating jacket

The electric heating jacket is mainly used as a heat source for reflow heating, and the heating temperature is above 300℃. It is a semi-circular (半圆的) inner cover woven with fiberglass wire and electric heating wire, with a metal shell on the outside and insulation material in the middle. According to the diameter of the inner sleeve, it can be divided into 50, 100, 150, 200, 250ml and other specifications, up to 3000ml. This equipment is not open flame heating, so it is safer to use. Because of its semicircular structure, the flask is in the hot air stream when heated, so the heating efficiency is higher. The outside drawing of electric heating jacket is shown in Figure 1-4.

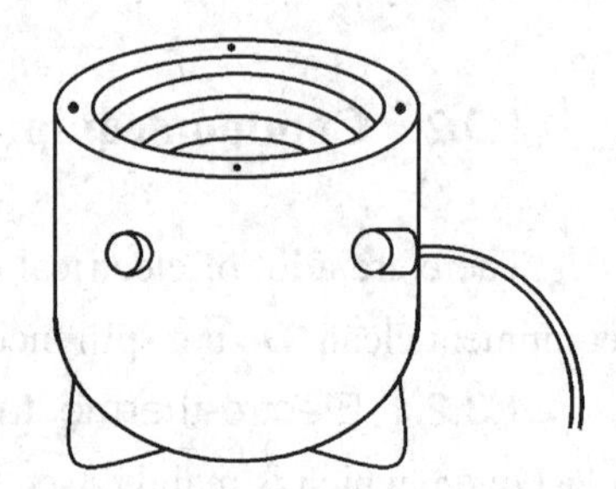

Figure 1-4 Electric Heating Jacket
图 1-4 电热套

Precautions for use: ① Do not sprinkle the medicine in the electric heating jacket, so as not to pollute the environment by volatilization of the medicine when heating, and to avoid the electric heating wire being corroded and broken. ② Put it in a dry place after use, otherwise the insulation performance will be reduced after internal absorption of moisture. ③ In case of moisture, pay attention to inductive electricity when using, and do not touch the inner core with hands. To slowly warm up, so that it can recover good insulation (绝缘) after drying.

1.3.2.4 Voltage regulating transformer Voltage regulator transformers are mainly divided into two types in laboratory. First it can be connected to an electric heating jacket to adjust the temperature, and the other it can be connected to an electric mixer to adjust the stirring rate. The two functions can also be concentrated on a single instrument, which is more convenient to use. However due to the different internal structures of the two voltage regulator transformers, the two instruments cannot be used interchangeably, otherwise the instruments will be burned. The outside drawing of voltage regulating transformer is shown in Figure 1-5.

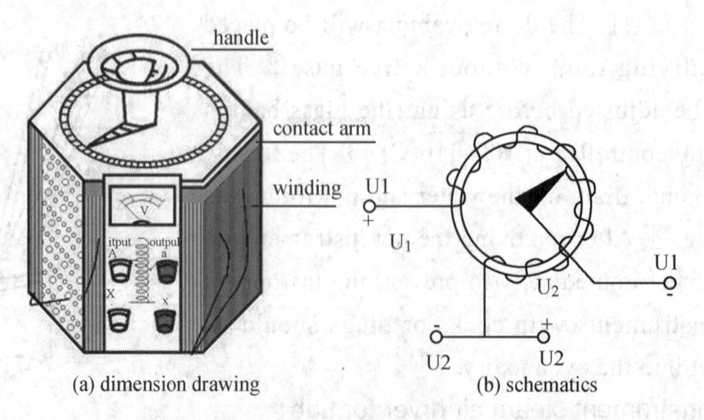

Figure 1-5 Voltage Regulating Transformer
图 1-5 调压变压器

Precautions for use: ① Adjust the voltage regulator to zero before turning on the power. ② When the old voltage (电压) regulator is used, pay attention to safety and connect the ground wire to prevent the case from being charged. Pay attention not to connect the output and input terminals incorrectly.

③ First connect to power supply when used, and then adjust the knob slowly to the desired position (adjust according to the heating temperature or stirring speed). No matter what kind of regulating transformer is used, overload operation is prohibited and maximum usage is 2/3 of full load. ④ After used, turn the knob to zero, turn off the switch, unplug the power plug and place it in a dry and ventilated place. Keep the voltage regulating transformer clean to prevent corrosion.

1.3.2.5 Stirrer The stirrer is generally used to stir liquid reactants during reaction. The agitator is divided into electric stirrer and electromagnetic stirrer, as shown in Figure 1-6.

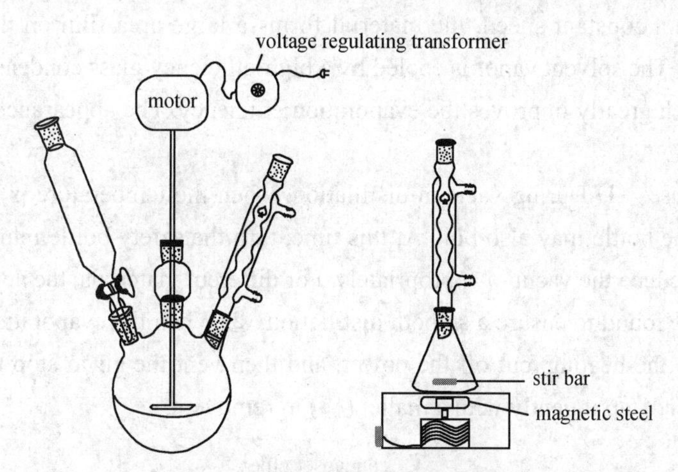

Figure 1-6 Electric Stirrer and Electromagnetic Stirrer
图 1-6 电动搅拌器与电磁搅拌器

(1) electric stirrer The stir bar should first connect with electric mixer, then the other side of the stir bar connect to the reaction bottle place with a sleeve or plug. Fix all the parts and components, then turn on the timer after power on and then turn on the power switch for work.

Precautions for use: ① Where the Erlenmeyer flask (锥形瓶) is used for mixing, align the mixing rod at the center and then start the machine for mixing. ② If the power is off suddenly during use, please check the power plug to see if there is loosening, also if found there is decantation during operation or unstable mixing, please stop the machine for checkup, adjust and screw down the clamp to guarantee the centration of mixing rod.

(2) electromagnetic stirrer The electromagnetic stirrer can be used in a completely sealed reaction. The motor drives the magnet to rotate, and the magnet drives the magnetic stir bar to rotate in the reactor, so as to achieve the purpose of stirring. Generally, the electromagnetic stirrer is equipped with a rotary knob for temperature and speed control. After use, the knob shall be returned to zero, and the power plug shall be pulled out. During use, attention shall be paid to moisture-proof and corrosion-resistant.

Use method: ① Place the reaction vessel containing the solution and stir bar on the workbench. ② Turn on the external power supply, turn on the power switch, and the indicator light (green light) will be on. ③ Select heating. Insert the sensor into the solution, adjust the temperature control knob to the required temperature scale, and the indicator light (red light) will be on, which means the heating state (when the red light goes out, it is in constant temperature state). ④ Gradually adjust the speed control knob to the required speed. It is strictly forbidden to start directly at high speed, which will cause the stir bar to be out of sync, then have the phenomenon of jumping. ⑤ After the work is completed, set the speed control and temperature control knob to the small position, turn off the power switch and cut off the power supply.

Precautions for use: ① The power supply shall be equipped with three pin safety socket and well grounded. ② All kinds of liquids are forbidden to enter the machine to avoid damaging the internal organs of the machine. ③ Vessels with flat bottom shall be selected. ④ After the completion of the work, the working panel and sensor shall be wiped clean, and there shall be no water drop and dirt residue on them.

1.3.2.6 Rotary evaporator Rotary evaporators (旋转蒸发仪) are mainly used for Liquid concentration and solvent recovery in the pharmaceutical, chemical and biopharmaceutical industries. It generally consists of a rotatable evaporator (usually a round bottom flask), a condensing tube and a receiving flask. The principle is that under vacuum conditions, constant temperature heating makes the rotary bottle rotate at a constant speed, the material forms a large area film on the bottle wall, and is efficiently evaporated. The solvent vapor is cooled by a high-efficiency glass condenser and recovered in a collection bottle, which greatly improves the evaporation efficiency. The appearance of rotary evaporator is shown in Figure 1-7.

Precautions for use: ① During vacuum distillation, when the temperature is high and the vacuum is low, the liquid in the bottle may also boil. At this time, turn the safety bottle intubation switch in time and let in cold air to reduce the vacuum appropriately. For different materials, the appropriate temperature and vacuum should be found to ensure a smooth distillation. ② When the evaporation stops, first stop the rotation in order, stop the heating, cut off the power, and then vent the air to stop the vacuum. ③ If the flask cannot be removed, tap it gently with a mallet (棒) to remove it.

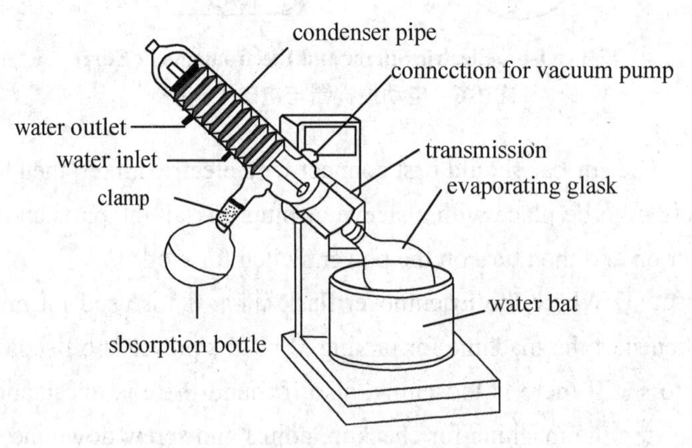

Figure 1-7 Rotary Evaporator

图 1-7 旋转蒸发仪

1.3.2.7 Electronic balance Electronic balance (Figure 1-8) is a common weighing equipment in laboratory, especially in micro and semi micro experiments. FA/JA series electronic balance features high precision and strong environment adaptation. As a series product, it can be widely applied to textile, petroleum, chemical industry, medicine and academies et al.

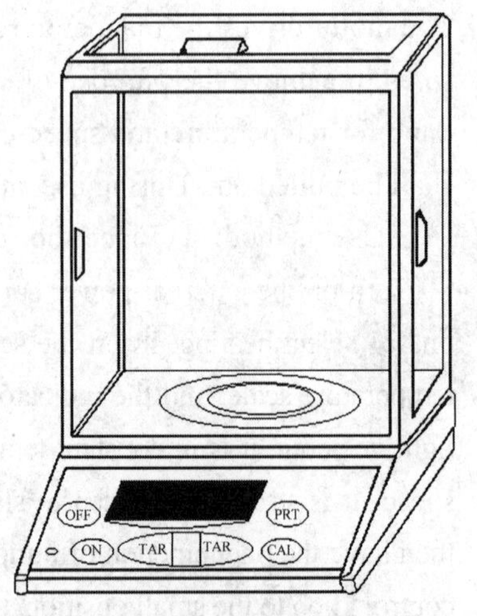

Figure 1-8 Electronic Balance

图 1-8 电子天平

Method of application: ① Place balance on the stable and level worktable. No vibration, drafts, direct sunlight or excessive temperature fluctuations (波动). ② Fix scale pan and adjust leveling feet to keep bubble at the center of bubble-level. ③ Confirm the current AC voltage whether complaint with the power supply adapter covered with the balance before connecting the power supply. ④ Press < ON > key in the weighing pan unloading. ⑤ To obtain

accurate weighing results, balance must be left switched on for at least 30 minutes before carrying on the first weighing operation. ⑥ To obtain accurate weighing results, the balance must be adjust to match the gravitational acceleration at its location. Adjusting is necessary after the balance preheating is finished. ⑦ The balance can weigh after calibrating. When weighing, the number can be read after "O" sign extinguishes at the lower left corner of indicator.

Precautions for use: ① The total weight of the container and the substances to be weighed must not be greater than the maximum weight range of the balance, otherwise the transducer of the balance can be damaged. ② In the weighing process, the substance weighed must be laid lightly and confirm the balance not overloaded to avoid damaging the transducer (传感器) of the balance.

1.3.2.8 Circulating water multi-purpose vacuum pump Circulating water multi-purpose vacuum pump is a new type of multi-purpose vacuum pump designed with circulating water as the fluid and producing negative pressure by jets. It is widely used in evaporation, distillation, crystallization, filtration, decompression, sublimation and other operations. Since water can be recycled, direct drainage is avoided, and the water saving effect is obvious. Therefore, it is the ideal decompression equipment in the laboratory. Water pumps are generally used in decompression systems that do not require high vacuum. Figure 1-9 is the appearance of SHB−Ⅲ type circulating water multi-purpose.

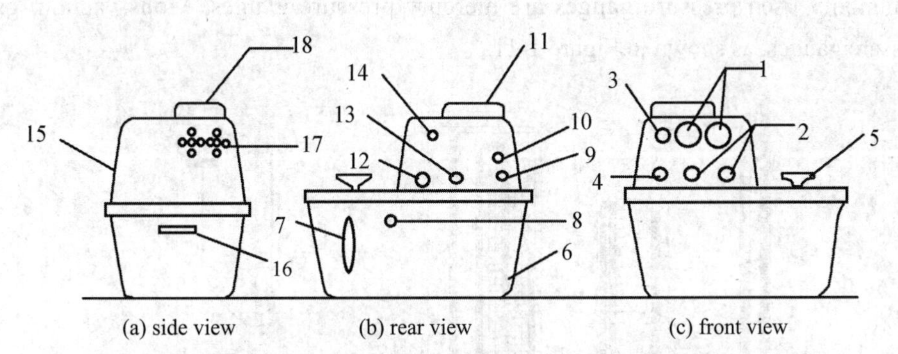

Figure 1-9 SHB-Ⅲ Type Circulating Water Multipurpose Vacuum Pump

图 1-9 循环水多功能真空泵

1.vacuum gauge, 2.exhaust port, 3.power indicator light, 4.power switch, 5.handle of water tank cover, 6.water tank, 7.water drain hose, 8.overflow mouth, 9.power line inlet, 10.insurance seat, 11.motor hood, 12.circulating water outlet, 13.circulating water inlet, 14.circulating water switch, 15.top cap, 16.water tank handle, 17.heat vent, 18.motor hood.

Precautions for use: ① It is better to connect a safety buffer bottle to the suction port of the vacuum pump, so as preventing water from being sucked into the reaction bottle when the pump is stopped, the reaction fails. ② Before turning on the pump, check whether it is connected to the system, and then open the cock on the safety buffer bottle. After turning on the pump, adjust the required vacuum with the cock. When closing the pump, first open the cock on the safety buffer bottle and put in air, remove the interface with the system, and then turn off the pump. Don't do the opposite. ③ The water in the pump should be replenished (补充) and replaced frequently to keep the pump clean and vacuum.

1.3.2.9 Oil pump The oil pump is also a common pressure reducing equipment in the laboratory, which is often used in the situation of high vacuum requirement. The efficiency of the pump depends on the mechanical structure of the pump and the quality of the vacuum pump oil (the lower vapor pressure of the oil is, the better efficiency is). A good vacuum pump can pump the vacuum degree below 100 Pa. The more precise the structure of the oil pump is, the higher the requirements for working conditions needs. Figure 1-10 is the schematic diagram of oil pump and protection system.

In the process of vacuum distillation with oil pump, solvent, water and acid gas will pollute the oil, which may lead to increase the vapor pressure of the oil, reduce the vacuum degree, and the gas may corrode the pump body. In order to protect the oil pump, pay attention to the following points: ① Regular inspection, regular oil change, moisture-proof and anti-corrosion. ② Place protection devices at the inlet of the pump, such as paraffin absorption tower (to absorb organic matters), silica gel or anhydrous calcium chloride absorption tower (to absorb water vapor), sodium hydroxide absorption tower (to absorb acid gas), and cooling well (to condense impurities).

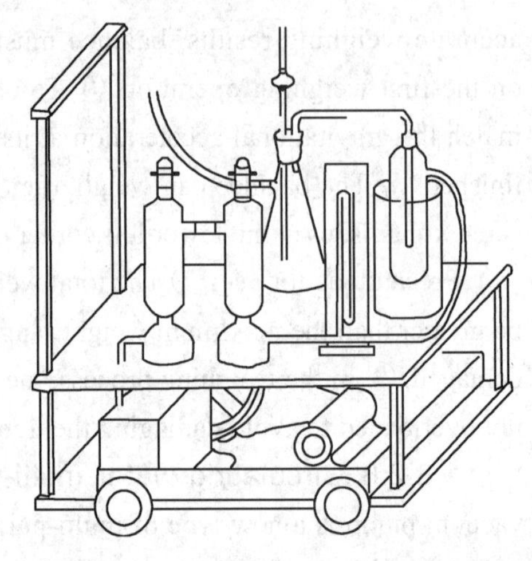

Figure 1-10 The Schematic Diagram of Oil Pump and Protection System

图 1-10 油泵及保护系统原理图

1.3.2.10 Vacuum pressure gauge Vacuum pressure gauges are often used in conjunction with water or oil pumps to measure the degree of vacuum in the system. Commonly used pressure gauges are mercury pressure gauges, Mohs vacuum gauges, and vacuum pressure gauges, as shown in Figure 1-11.

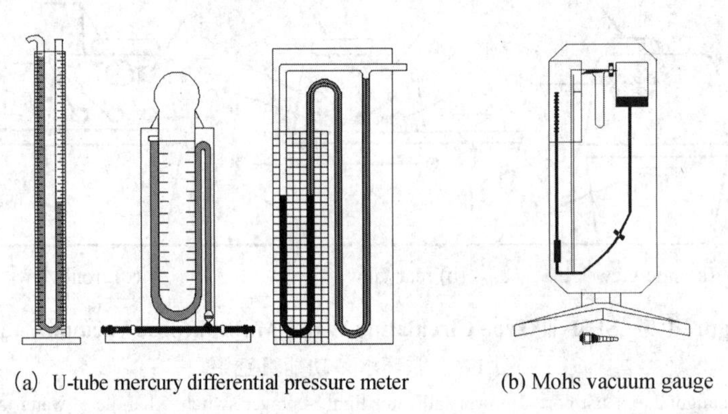

(a) U-tube mercury differential pressure meter (b) Mohs vacuum gauge

Figure 1-11 Pressure Gauge

图 1-11 压力计

Precautions for use: When you stop using the pump, slowly open the air release valve on the safety buffer bottle, and then turn off the pump. Otherwise, due to the high density of mercury (13.9g/cm^3), the glass tube will be broken during rapid flow, causing the mercury to spray out and causing pollution. When pulling and pushing the pump truck, care should be taken to protect the mercury pressure gauge from violent vibration.

1.3.2.11 Ultrasonic cleaner Ultrasonic (超声的) cleaners are commonly used to clean glass instruments in the laboratory. The principle of ultrasonic cleaning machine is to transform the sound energy of power ultrasonic frequency source into mechanical vibration through transducer, and to radiate the ultrasonic wave to the cleaning liquid in the tank through cleaning the tank wall. Due to the radiation of ultrasound, micro bubbles in the liquid in the tank can keep vibration under the action of sound wave. Destroy the adsorption between the dirt and the surface of the cleaning part, cause the fatigue damage to the dirt layer and be refuted, as shown in Figure 1-12. The vibration of the micro bubbles will scrub the solid surface.

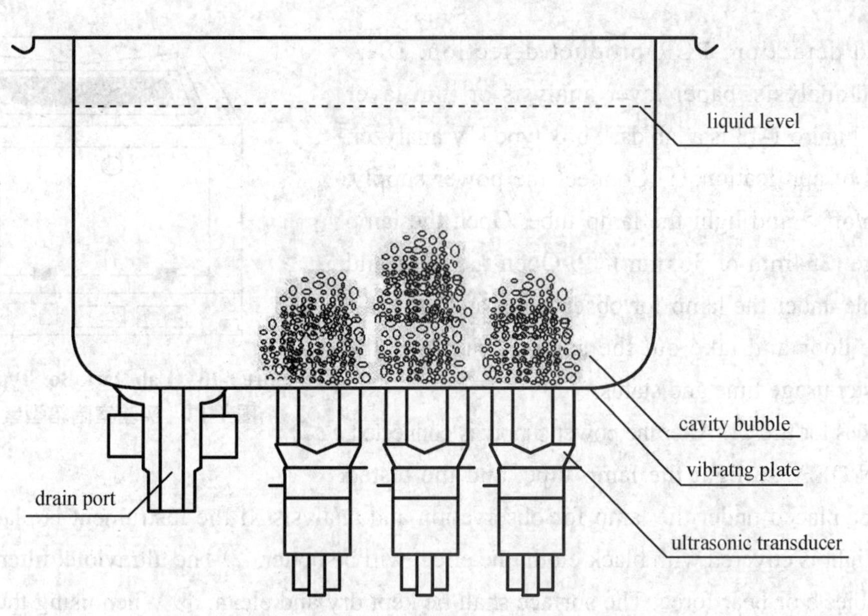

Figure 1-12 Schematic Diagram of Ultrasonic Cleaner
图 1-12 超声清洗器原理图

Method of application: ① Connect the ultrasonic cleaner to a 220V/50Hz three-core power supply. ② Put the objects to be cleaned into the cleaning rack and into the cleaning tank. ③ Injected water or aqueous solution reaches a certain water level. ④ Put the cleaning liquid into the cleaning tank proportionally. ⑤ Turn on the power and set the heating temperature at 0~80℃. ⑥ Set the ultrasonic time for 1~30 minutes. If you need to work for a long time, point the ultrasonic knob to ON. ⑦ Put the object to be cleaned into the cleaning rack and into the cleaning tank. Try not to put the object directly into the cleaning tank, so as not to affect the cleaning effect and damage the instrument. Then add the detergent proportionally. ⑧ After cleaning, remove the net frame from the cleaning tank, rinse with warm water, then dry and wipe with hot air. At the same time, put the uncleaned objects into the ultrasonic cleaning tank and press the ultrasonic key. ⑨ The solution in the cleaning tank can be used repeatedly. The service life can be decided according to the dirt condition of the object. When the emitted ultrasonic wave is saturated by the water solution, the sewage (污水) must be discharged. After cleaning, open the valve when the dirty liquid in the cleaning tank is discharged, and close the valve after exhausting.

Precautions for use: ① During the cleaning process, objects are put into the network rack and into the cleaning tank. Objects cannot be directly put into the bottom of the cleaning tank or collided with the bottom transducer. ② The use of appropriate cleaning chemicals must be compatible with the ultrasonic cleaning tank made of stainless steel, and the use of strong acid, strong alkali and volatile corrosive chemicals is not allowed. ③ If the aqueous solution is accidentally ingested or gets into the eye during the cleaning process, immediately rinse with plenty of water and seek medical treatment in time.

1.3.2.12 UV analyzer UV analyzer is the application of fluorescence (荧光) technology. When a substance is irradiated by an incident light of a certain wavelength, it enters the excited state after absorbing the light energy, and immediately de excites and emits an outgoing light with a longer wavelength than the incident light (usually the wavelength is in the visible light band); and once the incident light stops, the luminous phenomenon disappears immediately. The outgoing light with this property is called fluorescence. UV analyzers with different wavelengths are used in electrophoretic

analysis and detection, PCR product detection, DNA fingerprinted analysis, paper layer analysis or thin layer analysis, etc. Figure 1-13 is a lab dark box type UV analyzer.

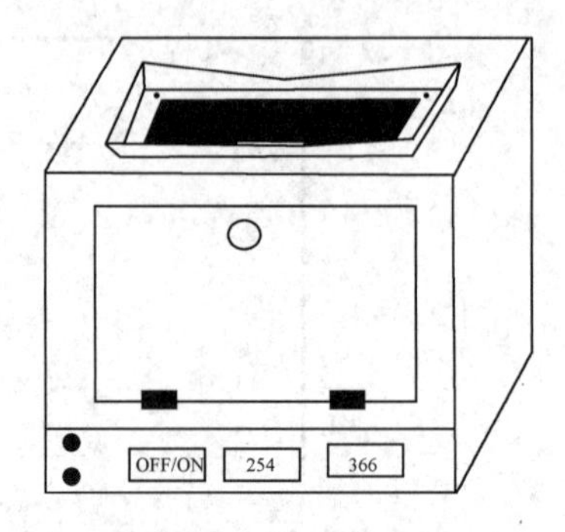

Figure 1-13 Lab Dark Box Type UV Analyzer
图 1-13 实验室暗箱型紫外分析仪

Method of application: ① Connect the power supply, turn on < on/off > and light the lamp tube. Open the lamp used in range (254mm or 365mm). ② Open the door and put the sample under the lamp for observation and analysis. ③ Open the door and take out the sample. Turn off the power. Register usage time and status.

Precautions for use: ① After the power supply is connected, turn on "ON/OFF" to light the lamp tube, and the tested sample can be placed under the lamp for observation and analysis. If the instrument is placed in a dark room or the light is covered with black cloth, the effect will be better. ② The ultraviolet filter shall not rub with metal objects or bear force. The surface shall be kept dry and clean. ③ When using the product, the operator should aim the ultra violet ray at the sample to avoid irradiating to the human body. It is best to wear protective glasses to avoid causing harm to the human body.

1.3.2.13 Low temperature coolant circulating pump The low temperature coolant (冷却剂) circulating pump is a low temperature liquid circulating equipment which adopts mechanical refrigeration, and it can provide low temperature liquid and low temperature water bath (Figure 1-14) . Combined with rotating evaporator, vacuum freeze-drying box, circulating water vacuum pump, magnetic stirrer and other instruments, the chemical reaction operation and drug storage under multi-functional low temperature are carried out.

Method of application: ① For the first use, open the top cover and add water or non-freezing liquid from the mouth of the tank. Add it to the top mouth along the inlet nozzle about 30mm lower, the refrigeration coil (or the inlet) must be submerged. ② The coolant circulation pump is connected to the system used in the experiment. ③ Switch on the power supply and turn on the protector switch. Press the "power" button, and the R indicator flashes. The display window displays room temperature. After the delay of 3 minutes, the R indicator light will be on, press the "refrigeration" button, the cooling machine will start

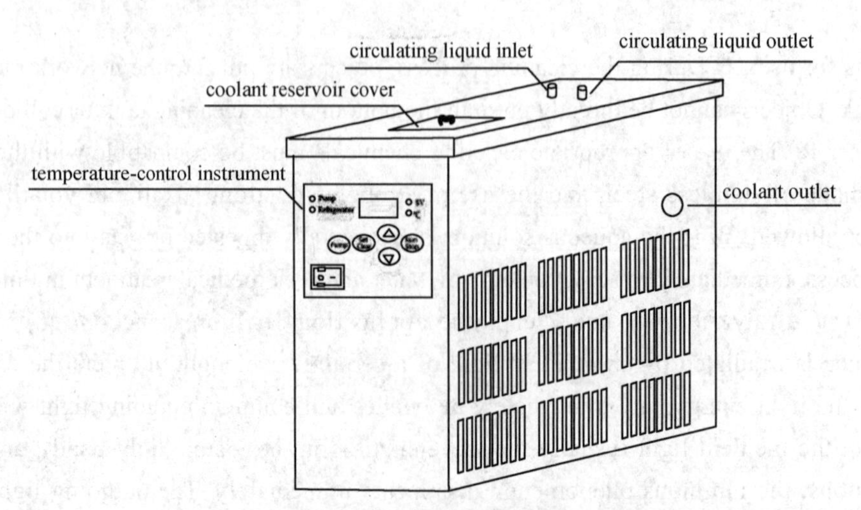

Figure 1-14 Low Temperature Coolant Circulating Pump
图 1-14 低温冷却循环泵

to work. ④ Set temperature required for the experiment.

Precautions for use: ① The power cord should be reliably grounded ! ② If there is no circulating coolant in the liquid container or the liquid level is too low, the circulating pump cannot be started. ③ There should be enough clear space around the cryogenic (冷冻的) coolant circulating pump. ④ There is a high pressure gas in the closed refrigeration system, non-professional personnel do not open maintenance.

三、有机化学实验室常用仪器

（一）玻璃仪器及玻璃仪器的清洗、干燥

1. 常用玻璃仪器 玻璃仪器一般分为普通玻璃仪器和标准磨口玻璃仪器两种。在实验室常用的普通玻璃仪器有锥形瓶、烧杯、布氏漏斗、吸滤瓶、普通漏斗、分液漏斗等，见图 1-1（a）。常用的标准磨口仪器有圆底烧瓶、三口瓶、蒸馏头、冷凝器、接收管等，见图 1-1（b）。玻璃仪器的用途见表 1-1。

表 1-1 有机化学实验常用仪器的应用范围

仪器名称	应用范围	备注
圆底烧瓶	用于反应、加热回流及蒸馏等	
三口圆底烧瓶	用于反应，三口分别安装电动搅拌器、回流冷凝管及温度计等	
直形、空气冷凝管	用于蒸馏	
球形（或蛇形）冷凝管	用于加热回流	
蒸馏头	与圆底烧瓶组装后用于蒸馏	
单股接收管（单尾接液管）	用于常压蒸馏	
双股接收管（多尾接液管）	用于减压蒸馏	
分馏柱	用于分馏多组分混合物	
恒压滴液漏斗	用于内有压力的反应体系的液体滴加	
分液漏斗	用于溶液的萃取及分离	
锥形瓶	用于存放液体、混合溶液及加热小量溶液	不能用于减压蒸馏
烧杯	用于加热溶液、浓缩溶液及用于溶液混合和转移	
量筒	量取液体	切勿用于加热
吸滤瓶	用于减压过滤	切勿用于加热
布氏漏斗	用于减压过滤	瓷质
瓷板漏斗	用于减压过滤	瓷质，瓷质板为活动圆孔板
熔点管	用于测熔点	内装液状石蜡、硅油或浓硫酸等
干燥管	装干燥剂，用于无水反应装置	

标准磨口仪器口径的大小通常用数字编号来表示，该数字是指磨口最大端直径的毫米整数，根据磨口口径分为 10、14、19、24、29、34、40、50 等型号。相同编号的子口与母口可以连接。当用不同编号的子口磨口与母口连接时，中间可加上一个大小口接头。有时也用两组数字来表示

口径的大小及长度，例如 14/30，表示仪器的磨口直径为 14mm，磨口长度为 30mm。学生使用的常量仪器一般是 19 号或 24 口的磨口仪器，半微量实验中采用的是 14 号的磨口仪器，微量实验中采用 10 号磨口仪器。使用玻璃仪器时应注意以下几点。

（1）使用时，应轻拿轻放。

（2）不能用明火直接加热玻璃仪器，加热时应垫石棉垫。

（3）不能用高温加热不耐温热的玻璃仪器，如吸滤瓶、普通漏斗、量筒等。

（4）玻璃仪器使用完后，应及时清洗干净，特别是标准磨口仪器放置时间太久，容易黏结在一起，很难拆开。如果发生此情况，可用热水烫黏结处或用热风吹母磨口处，使其膨胀而脱落，还可用木槌轻轻敲打黏结处使之脱落。洗干净的玻璃仪器最好自然晾干。

（5）带旋塞或具塞的仪器清洗后，应在塞子和磨口接触处夹放纸片或涂抹凡士林，以防黏结。

（6）标准磨口仪器磨口处要干净，不得粘有固体物质。清洗时，应避免用去污粉擦洗磨口。否则，会使磨口连接不紧密，甚至会损坏磨口。

（7）安装仪器时，应做到横平竖直，磨口连接处不应受歪斜的应力，以免仪器破裂。

（8）使用玻璃仪器时，磨口处一般无需涂润滑剂，以免粘附反应物或产物。但是反应中使用强碱时，则要涂润滑剂，以免磨口连接处因碱腐蚀而黏结在一起，无法拆开。当减压蒸馏时，应在磨口连接处涂润滑剂，保证装置密封性良好。

（9）使用温度计时，应注意不要用冷水冲洗热的温度计，以免破裂，尤其是水银球部位，应冷却至近室温后再冲洗。切记不能将温度计当作玻棒用于搅拌，以免温度计损坏。

2. 实验室玻璃仪器的清洗原则 使用后立即进行清洁，通常容易清洁玻璃仪器。使用清洁剂时，通常是用于实验室玻璃仪器的清洁剂，例如 Liquinox 或 Alconox。通常，既不需要也不希望使用清洁剂或自来水。你可以用适当的溶剂冲洗玻璃仪器，然后用蒸馏水冲洗几次，最后用去离子水冲洗。

（1）普通化学品的洗涤　水溶性溶液（例如氯化钠或蔗糖溶液）：用去离子水冲洗 3~4 次，然后将玻璃仪器置于仪器架上自然沥干。

水不溶性溶液（例如己烷或氯仿溶液）：用乙醇或丙酮冲洗 2~3 次，用去离子水冲洗 3~4 次，然后将玻璃仪器置于仪器架上自然沥干。在某些情况下，初次冲洗需要使用其他溶剂。

强酸（例如浓 HCl 或 H_2SO_4）：在通风橱下，用大量自来水小心地冲洗玻璃仪器，再用去离子水冲洗 3~4 次，然后将玻璃仪器置于仪器架上自然沥干。

强碱（例如 6mol/L NaOH 或浓 NH_4OH）：在通风橱下，用大量自来水小心地冲洗玻璃仪器，再用去离子水冲洗 3~4 次，然后将玻璃仪器置于仪器架上自然沥干。

弱酸（例如醋酸溶液和 0.1mol/L 或 1mol/L HCl、H_2SO_4）：用去离子水冲洗 3~4 次，然后将玻璃仪器置于仪器架上自然沥干。

弱碱（例如 0.1mol/L、1mol/L NaOH 和 NH_4OH）：用自来水彻底冲洗以除去碱，再用去离子水冲洗 3~4 次，然后将玻璃仪器置于仪器架上自然沥干。

（2）特殊玻璃仪器的洗涤　用适当的溶剂冲洗玻璃仪器，再用去离子水洗涤水溶性物质。使用乙醇洗涤可除去溶于乙醇的内容物，然后用去离子水冲洗。也可根据需要先用其他溶剂冲洗，然后用乙醇冲洗，最后用去离子水冲洗。如果玻璃仪器需要擦洗，请使用热肥皂水用刷子擦洗，然后用自来水彻底冲洗，最后再用去离子水冲洗。

滴定管的洗涤：先用热肥皂水洗涤，再用自来水彻底冲洗，然后用去离子水冲洗 3~4 次。确保最后的冲洗液从滴定管的尖嘴部位流出。滴定管需要彻底清洁才能用于实验室的定量分析。

移液管和容量瓶：在特定的情况下，可将玻璃仪器浸泡在肥皂水中过夜。然后使用温和的肥皂水清洁移液管和容量瓶。玻璃仪器可能需要用刷子擦洗。用自来水冲洗，再用去离子水冲洗三

至四次。

3. 干燥或不干燥

（1）不干燥玻璃仪器　如果你要向玻璃仪器中加水，可以将其保持湿润（除非会影响最终溶液的浓度）。如果最终使用的玻璃仪器中溶液的浓度受水的影响，我们需要使用标准溶液润洗三次。

（2）干燥玻璃仪器　不建议用纸巾或强制通风的方式来干燥玻璃仪器，因为这可能会引入污染溶液的纤维或杂质。通常可以让玻璃仪器在架子上风干。

如果要在清洗后立即使用玻璃仪器，并且必须将其干燥，可使用丙酮冲洗 2~3 次。这将清除所有水分并迅速蒸发。将空气吹入玻璃仪器以使其干燥不是一个好方法，但可以使用真空干燥以蒸发溶剂。

4. 其他注意事项　不使用玻璃仪器时，取下塞子和活塞。否则，它们可能会在里面“冻结”，不易取出。

使用浸有乙醚或丙酮的无绒毛巾擦拭玻璃接缝处的油脂。戴上手套并避免吸入挥发的液体。

使用干净的玻璃仪器倾倒去离子水冲洗液时，应形成平滑的液床。如果看不到这种现象，则可能需要更有效的清洁方法。

（二）常用设备、配套仪器及使用

实验室有很多电器设备，使用时应注意安全，并保持这些设备的清洁，千万不要将药品洒到设备上。

1. 烘箱　实验室一般使用的是恒温鼓风干燥箱（图 1-2），主要用于干燥玻璃仪器或无腐蚀性、热稳定好的药品。

使用注意：①干燥箱将放置在无活性气体地面水平的干燥室。②使用时应先调好温度 (烘玻璃仪器一般控制在100~110℃)。③刚洗好的仪器应将水倒尽沥干，无水滴下后再放入烘箱中。④热仪器取出后，不要马上碰冷的物体如冷水、金属用具等带旋塞或具塞的仪器，取下塞子后再放入烘箱中烘干。

2. 气流烘干器　气流烘干器是一种用于快速烘干仪器的设备，如试管、烧杯等的干燥。实验室用气流烘干器如图 1-3 所示。

使用注意：使用时，将仪器洗干净后，沥干水分，然后将仪器套在烘干器的多孔金属管上。先吹冷风、然后吹热风，最后再用冷风吹，使玻璃仪器冷却至室温，不使水汽再冷凝在容器内。气流烘干器不宜长时间加热，以免烧坏电机和电热丝。

3. 电热套　电热套主要用做回流加热的热源，加热温度高达 300℃ 以上。它是用玻璃纤维丝与电热丝编织成半圆形的内套，外边加上金属外壳，中间填上保温材料，如图 1-4。根据内套直径的大小分为 50、100、150、200、250ml 等规格，最大可到 3000ml。此设备不是明火加热，使用较安全。由于它的结构是半圆形的，在加热时，烧瓶处于热气流中，因此，加热效率较高。

使用注意：①不要将药品洒在电热套中，以免加热时药品挥发污染环境，同时避免电热丝被腐蚀而断开。②用完后放在干燥处，否则内部吸潮后会降低绝缘性能。

如遇潮湿，使用时注意感应电，不要用手触摸内芯。慢慢预热，使其干燥后能恢复良好的绝缘性。

4. 调压变压器　调压变压器主要分为两类。一类可与电热套相连用来调节电热套温度；另一类可与电动搅拌器相连用来调节搅拌器速度。也可以将两种功能集中在一台仪器上，这样使用起来更为方便。但是两种仪器由于内部结构不同不能相互串用，否则会将仪器烧毁。调压变压器外形图如图 1-5 所示。

使用注意：①先将调压器调至零点，再接通电源。②使用旧式调压器时，应注意安全，要接好地线，以防外壳带电。注意输出端与输入端不要接错。③使用时，先接通电源，再调节旋钮到所需要的位置(根据加热温度或搅拌速度来调节)。调节变换时，应缓慢进行。无论使用哪种调压变压器都不能超负荷运行，最大使用量为满负荷的2/3。④用完后将旋钮调至零点，关上开关，拔掉电源插头，放在干燥通风处，应保持调压变压器的清洁，以防腐蚀。

5. 搅拌器　一般用于反应时搅拌液体反应物，搅拌器分为电动搅拌器和电磁搅拌器，如图1-6。

（1）电动搅拌器　搅拌棒应先与电动搅拌机连接，然后用套筒或塞子将搅拌棒的另一侧连接到反应瓶处。安装好所有零部件，通电后打开定时器，然后打开电源开关工作。

使用注意：①在使用锥形烧瓶进行混合的时候，将搅拌棒对准中心，然后启动机器进行混合。②使用过程中如突然断电，请检查电源插头有无松动，运行中如发现偏心或搅拌不稳，请停机检查，调整并拧紧卡箍，以保证搅拌棒处于中心位置。

（2）电磁搅拌器　电磁搅拌器能在完全密封的装置中进行搅拌。它由电机带动磁体旋转，磁体又带动反应器中的磁子旋转，从而达到搅拌的目的。电磁搅拌器一般都带有温度和速度控制旋转钮，使用后应将旋钮回零，拔掉电源插头，使用时应注意防潮防腐。

使用方法：①将装有溶液的反应容器和搅拌棒放在工作台上。②接通外部电源，打开电源开关，指示灯（绿灯）亮。③选择加热，将传感器插入溶液中，将温度控制旋钮调整到所需的温度刻度，指示灯（红灯）亮，表示加热状态。红灯熄灭时，处于恒温状态。④逐渐将速度控制旋钮调整到所需的速度。严禁高速直接启动，否则会造成搅拌子不同步，进而出现跳跃现象。⑤工作完成后，将速度控制和温度控制旋钮调到小位置，关闭电源开关，切断电源。

使用注意：①电源需配置三孔安全插座且接地良好。②禁止各种液体进入仪器内部，以免损坏机器内部零件。③选用平底容器。④工作结束后，应将工作面板和传感器擦拭干净，不得有水滴和污垢残留。

6. 旋转式蒸发器　旋转式蒸发器主要用于制药、化工、生物制药等行业的液体浓缩和溶剂回收。它通常由一个可旋转的蒸发器(通常是一个圆底烧瓶)、一个冷凝管和一个接收烧瓶组成。此装置可在常压或减压下使用，可一次进料，也可分批进料。其原理是在真空条件下，恒温加热使旋转瓶匀速旋转，物料在瓶壁上形成大面积薄膜，并有效蒸发。溶剂蒸汽由高效的玻璃冷凝器冷却，回收到收集瓶中，大大提高了蒸发效率。旋转式蒸发器如图 1-7 所示。

使用注意：①减压蒸馏时，当温度高、真空度低时，瓶内液体也可能会暴沸。此时，及时转动安全瓶插管开关，通入冷空气适当降低真空度即可。对于不同的物料，应找出合适的温度与真空度，确保平稳地进行蒸馏。②停止蒸发时，先依次停止旋转，停止加热，切断电源，再通入空气，停止抽真空。③若烧瓶取不下来，可趁热用木槌轻轻敲打，以便取下。

7. 电子天平　电子天平是实验室中常见的称重设备，尤其是在微量和半微量实验中。FA/JA系列电子天平具有高精度和强环境适应性，如图1-8。作为系列产品，在纺织、石油、化工、医药、科学研究等领域广泛应用。

使用方法：①将天平放在平稳的工作台上。无振动、气流、阳光直射或大的温度波动。②固定秤盘并调节，以使气泡保持在气泡水平仪的中心。③连接电源之前，确认当前的交流电压是否符合天平的电源适配器。④按下秤盘上的 <ON> 键。⑤为了获得准确的称量结果，在进行第一次称量操作之前，必须将天平至少打开 30 分钟。⑥为了获得准确的称重结果，必须调整天平，使其与所在位置的重力加速度相匹配。天平预热结束后需要调整。⑦天平标定后可以称重。称重时，可在指示器左下角的“O”符号熄灭后读取数字。

使用注意：①容器和被称量物质的总重量不得超过天平的最大重量范围，否则会损坏天平的传感器。②在称量过程中，必须轻轻地放置被称量的物质，并确保天平没有超载，以免损坏天平的传感器。

8. 循环水多用真空泵 循环水多用真空泵是以循环水作为流体，利用射流产生负压的原理而设计的一种新型多用真空泵，广泛用于蒸发、蒸馏、结晶、过滤、减压、升华等操作中。由于水可以循环使用，避免了直排水，节水效果明显。因此，是实验室理想的减压设备。水泵一般用于对真空度要求不高的减压体系中。图 1-9 为 SHB-Ⅲ型循环水多用真空泵的外观示意图。

使用注意：①真空泵抽气口最好接一个安全缓冲瓶，以免停泵时，水被倒吸入反应瓶中，使反应失败。②开泵前，应检查是否与体系接好，然后，打开安全缓冲瓶上的旋塞。开泵后，用塞调至所需要的真空度。关泵时，先打开安全缓冲瓶上的旋塞放进空气，拆掉与体系的接口，再关泵。切忌相反操作。③应经常补充和更换水泵中的水，以保持水泵的清洁和真空度。

9. 油泵 油泵也是实验室常用的减压设备。油泵常在对真空度要求较高的场合下使用。油泵的效能取决于泵的机械结构及真空泵油的好坏（油的蒸气压越低越好），好的真空油泵能抽到 100Pa 以下的真空度。油泵的结构越精密，对工作条件要求就越高。图 1-10 为油泵及保护系统示意图。

在用油泵进行减压蒸馏时，溶剂、水和酸性气体会造成对油的污染，使油的蒸气压增加，降低真空度，同时这些气体可能腐蚀泵体。为了保护油泵，使用时应注意做到：① 定期检查，定期换油，防潮、防腐蚀。②在泵的进口处放置保护装置，如石蜡片吸收塔（吸收有机物）、硅胶或无水氯化钙吸收塔（吸收水汽）、氢氧化钠吸收塔（吸收酸性气体）和冷却阱（冷凝杂质）。

10. 真空压力表 真空压力表常用来与水泵或油泵连在一起使用，用于测量体系内的真空度。常用的压力表有水银压力计、莫氏真空规、真空压力表，见图 1-11。

使用水银压力计时应注意：停泵时，先慢慢打开安全缓冲瓶上的放空气阀，再关泵。否则，由于汞的密度较大（$13.9g/cm^3$），在快速流动时，会冲破玻璃管，使汞喷出，造成污染。在拉出和推进推动泵车时，应注意保护水银压力计，防止剧烈振动。

11. 超声波清洗机 超声波清洗机常用于实验室玻璃仪器的清洗。超声波清洗机的原理是通过换能器，将功率超声频源的声能转换成机械振动，通过清洗槽壁将超声波辐射到槽子中的清洗液。由于受到超声波的辐射，使槽内液体中的微气泡能够在声波的作用下保持振动，如图 1-12 所示。破坏污物与清洗件表面的吸附，引起污物层的疲劳破坏而被剥离，气体型气泡的振动对固体表面进行擦洗。

使用方法：① 将超声波清洗机连接到一个 220V/50Hz 的三孔电源。② 将要清洗的物品放入清洗架和清洗槽。③ 注入水或水溶液达到一定水位。④ 将清洗液按比例放入清洗槽。⑤ 打开电源，将加热温度设定在 0~80℃。⑥ 设置超声波时间 1~30 分钟。如果您需要长时间工作，请将超声波旋钮指向“开”。⑦ 清洁后，从清洁槽中取出网框，用温水冲洗，然后干燥并用热空气擦拭。同时，将未清洗的物品放入超声波清洗槽，按超声波键。⑧ 清洗槽中的溶液可以重复使用。使用寿命可根据物体的污垢情况而定。当发出的超声波被水溶液饱和时，污水必须排出。清洗结束后，当清洗槽内的脏液排出时，打开阀门，排出后关闭阀门。

使用注意：① 在清洗过程中，物体被放入网架和清洗槽中。不能将物体直接放入清洗槽底部或与底部传感器碰撞。② 使用合适的清洗剂必须与不锈钢制成的超声波清洗槽兼容，不允许使用强酸、强碱和易挥发的腐蚀性化学品。③如果在清洁过程中意外摄入清洗溶液或进入眼睛，立即用大量水冲洗并及时就医。

12. 紫外分析仪 紫外分析仪是荧光技术的应用。当某种常温物质经某种波长的入射光（通常是紫外线或 X 射线）照射，吸收光能后进入激发态，并且立即退激发并发出比入射光的波长长的出射光（通常波长在可见光波段）；而且一旦停止入射光，发光现象也随之立即消失。具有这种性质的出射光就被称之为荧光。不同波长的紫外分析仪用于电泳分析检测、PCR 产物检测、DNA 指纹分析、纸层分析或薄层分析等。图 1-13 为实验室暗箱型紫外分析仪。

使用方法： ①接通电源，按压 <on/off>，并打开范围（254mm 或 365mm）内使用的灯。②打

开箱门，把样品放在灯下观察分析。③打开箱门取出样品。关掉电源。

使用注意：①电源连接后，按压 <ON/OFF>，打开灯管，将被测样品置于灯下观察分析。如果把仪器放在暗室里或光线用黑布覆盖，效果会更好。②紫外线滤光片不得与金属物体摩擦或受力。滤光片表面应保持干燥和清洁。③使用紫外分析仪时，操作人员应将紫外线对准样品，避免对人体产生辐射。最好戴上防护眼镜，以免对人体造成伤害。

13. 低温冷却循环泵 低温冷却循环泵是采用机械制冷的低温液体循环设备，可提供低温液体和低温水浴，如图 1-14。常结合旋转蒸发器、真空冷冻干燥箱、循环水真空泵、磁力搅拌器等仪器，进行多功能低温化学反应操作和药品储存。

使用方法：①首次使用时，打开顶盖，从冷却液槽口加入水或冷却液。将其沿入口喷嘴下约 30mm 处添加到顶部，制冷盘管（或入口）必须浸没。②冷却循环泵与实验中使用的反应系统结合使用。③打开电源并打开保护开关。按下“Power”按钮，R 指示灯闪烁。显示窗口显示室温。延时 3 分钟后，R 指示灯亮，按下“Refrigeration”按钮，制冷机开始工作。④设定实验所需的温度。

使用注意：①电源线应接地！②如果液箱中没有循环冷却液或液位过低，不能启动循环泵。③低温冷却循环泵周围应有足够的净空间。④封闭式制冷系统中有高压气体，非专业人员不可打开维修。

1.4 Common Organic Solvent

Physical and chemical constants of common organic solvent are the important parameters for students to carry out experiments safely and successfully, which shall be helpful to improve students' understanding of the experimentation.

According to the specifications of the experiment for solvents, commercially available organic solvents are usually classified as analytical reagents (AR), chemical pure reagents (CP), industrial reagents (CD), etc, and the purification is generally not required. In the following cases, the solvents should be purified; the extremely high purity of solvent is need and the commercial solvent cannot meet the demand; the increase of impurities in a long-stalled solvent will hinder its use due to oxidation, moisture absorption and light, etc; replacing the higher-specification solvent with the lower-specification to reduce the cost in overuse of a solvent; reuse and recycle of solvents.

1.4.1 Methyl Alcohol

Additional names: Methanol, Carbinol; Molecular formula: CH_3OH; Molecular weight: 32.04; Elemental analysis: C 37.48%, H 12.58%, O 49.93%.

Properties: Boiling point 64.96℃; Melting point −97.81℃; Relative density (d_4^{20}) 0.792; Refractive index (n_D^{20}) 1.3292; Flash point 12℃. Explosive limit 5.5%~44.0% (V/V); Dielectric constant (ε) 32.7. Methanol is a colorless transparent flammable liquid that can be mixed with water in any proportion without forming an azeotropic (共沸的) mixture. Miscible with ethanol, ether, benzene, ketones and most other organic solvents. Slight alcoholic odor, crude material may have a repulsive, pungent odor. Burns with non-luminous, bluish flame. Caution: Poisoning may occur from ingestion, inhalation or percutaneous absorption.

Acute effects: Headache, fatigue, nausea, visual impairment or complete blindness (may be permanent), acidosis (酸中毒), convulsions (抽搐), mydriasis (瞳孔放大), circulatory collapse, respiratory failure, death.

Methanol with water content less than 0.5% can be removed by redistilling. Methanol with less than 0.01% water is dried by fractionation or 4Å molecular sieve.

1.4.2 Ethanol

Additional names: Absolute alcohol, Anhydrous alcohol, Dehydrated alcohol, Ethyl hydrate; Molecular formula: C_2H_5OH; Molecular weight: 46.07; Elemental analysis: C 52.14%, H 13.13%, O 34.73%.

Properties: Boling point 78.50℃; Melting point −114.10℃; Relative density (d_4^{20}) 0.789; Refractive index (n_D^{20}) 1.3292; Flash point 12℃; Explosive limit 3.3%~19.0% (V/V).

Clear, colorless, very mobile, flammable liquid; pleasant odor; burning taste; absorbs water rapidly from air. Miscible with water, ether, benzene, petroleum ether and other organic solvents. An azeotropic mixture with a constant boiling point 78.17℃ and a water content 4.47% (W/W) can be formed. Industrial alcohol contains a small amount of methanol. Medical alcohol mainly refers to about 75%

ethanol aq, also including other concentrations of ethanol that are widely used in medicine.

Common impurities are water, acetone, formaldehyde, etc, and commercial anhydrous ethanol often contains benzene, toluene, etc. Caution: Keep tightly closed, cool, and away from flame!Ethanol and methyl ether are isomers of each other.

1.4.3 n-Propanol

Additional names: 1-Propyl alcohol, Propanol; Molecular formula: C_3H_7OH; Molecular weight: 60.10; Elemental analysis: C 59.96%, H 13.42%, O 26.62%.

Properties: Boiling point 97.2℃; Melting point −126.5℃; Relative density (d_4^{20}) 0.804; Flash point 15℃. Explosive limit 2.0%~13.3%(V/V). n-Propanol is a colorless, transparent liquid with an ethanol-like odor, soluble in water, ethanol and ethyl ether. An azeotropic mixture with a constant boiling point 87℃ and a water content 28.3% (W/W) can be formed. It is mainly used as a fungicide (杀菌剂) for fuel oil, pesticides and pharmaceutical raw materials, spice materials, etc.

1.4.4 Ethyl Acetate

Additional names: Acetic acid ethyl ester, Acetic ether, EtOAc; Molecular formula: $CH_3COOC_2H_5$; Molecular weight: 88.11; Elemental analysis: C 54.53%, H 9.15%, O 36.32%.

Properties: Boiling point 77℃; Melting point −84℃; Relative density (d_4^{20}) 0.902; Relative density (d_4^{20}) 1.3719; Flash point −4℃; Dielectric constant (ε) 6.0; Explosive limit 2.2%~11.2% (V/V). Clear, volatile, flammable liquid, characteristic fruity odor; Pleasant taste when diluted. Slowly decomposed by moisture, then acquires an acid reaction. Absorbs water (up to 3.3% W/W). Miscible with ethanol, acetone, chloroform and ether, soluble in water (10% V/V). EtOAc can also dissolve some metal salts, such as lithium chloride, cobalt chloride, zinc chloride and ferric chloride, etc. Its vapor is heavier than air so that it spreads far enough away to catch fire in the lower. Keep tightly closed in a cool place and away from fire.

Common impurities are water, ethanol and acetic acid. The acid can be removed by washing once or twice with 5% sodium carbonate aq. Also, the alcohol can be washed off with saturated calcium chloride solution, and then dried with anhydrous calcium chloride.

1.4.5 Hexane

Additional names: Normal hexane, n-Hexane; Molecular formula: $C_6H_{14;}$ Molecular weight: 86.18; Elemental analysis: C 83.63%, H 16.37%.

Properties: Boiling point 68.74℃; Melting point −95.30℃; Relative density (d_4^{20}) 0.692; Refractive index (n_D^{20}) 1.388; Flash point 30° F (−1.11℃). Colorless, flammable liquid; faint, peculiar odor. Insoluble in water; miscible with acetone, chloroform, ether. The solubility ratio of n-hexane to ethanol is 50: 100 (33℃). As a good non-polar organic solvent, n-hexane is widely used in chemical industry and mechanical equipment surface cleaning.

Caution: Poisonous! Potential symptoms of overexposure are light-headedness, nausea, headache, numbness of extremities (四肢麻木), muscle weakness, irritation of eyes and nose, dermatitis (皮炎); chemical pneumonia, giddiness (眩晕).

1.4.6 Cyclohexane

Additional name: Hexamethylene; Molecular formula: C_6H_{12}; Molecular weight: 84.16; Elemental analysis: C 85.63%, H 14.37%.

Properties: Boiling point 80.7℃; Freezing point 6.5℃; Relative density (d_4^{20}) 0.778; Refractive index (n_D^{20}) 1.4266; Flash point −16.5℃. Explosive limit 1.3%~8.3% (V/V). Flammable liquid. Solvent odor. Pungent when impure, Its vapor is heavier than air so that it spreads far enough away to catch fire in the lower. Insoluble in water; miscible with ethanol, ethyl ether, benzene, acetone. The solubility

ratio of cyclohexane to methanol is 57 : 100 (25°C). It can be used as a standard substance/solvent for chromatographic analysis, resin, coating and paraffin oil. It can also be used to prepare cyclohexanol and cyclohexanone.

1.4.7 Petroleum Ether

Additional names: Petroleum benzin, Ligroin; Molecular formula: C_5H_{12}, C_6H_{14}, C_7H_{16}; Molecular weight: A mainly mixtures of pentane and hexane; Elemental analysis: About C 83.63%, H 16.37%.

Properties: Boiling point 30~80°C; Boiling range 30~60°C, 60~90°C, 90~120°C; Melting point < –73°C; Relative density (d_4^{20}) 0.64~0.66; Flash point <–20°C; Explosive limit 1.1%~8.7%(V/V). A colorless, transparent liquid with a smell of kerosene. It's basically a mixture of pentane and hexane with weak polar. Insoluble in water, soluble in anhydrous ethanol, benzene, chloroform, oil and other organic solvents. It is flammable and explosive, and can react strongly with oxidizer. Its vapor is heavier than air so that it spreads far enough away to catch fire in the lower. It is mainly used as solvent and grease treatment.

It is usually obtained from platinum-reforming residual oil or straight-run gasoline by fractionation and hydrogenation or other methods. Petroleum ether is different from gasoline, and there is no ether bond (C-O-C) in its structure. It can cause peripheral neuritis (周围神经炎) as well as a have strong irritant to the skin.

1.4.8 Ethyl Ether

Additional names: Ethoxyethane ether, Diethyl ether, Ethyl oxide, Diethyl oxide; Molecular formula: $C_2H_5OC_2H_5$; Molecular weight: 74.12; Elemental analysis: C 64.82%, H 13.60%, O 21.59%.

Properties: Boiling point 34.6°C; Melting point –116.3°C; Relative density (d_4^{20}) 0.714; Flash point –45°C; Explosive limit 1.85%~36.5%(V/V). Mobile, very volatile, highly flammable liquid; Explosive! Characteristic odor, sweetish, pungent odor. Slightly soluble in water. Miscible with lower aliphatic alcohols, chloroform, benzene. Evaporation to dryness is not allowed, and never bring an open flame to distill ether. Tends to form explosive peroxide under the influence of air and light, esp. when evaporation to dryness is attempted. Peroxides may be removed from ether by shaking with 5% ferrous sulfate aq. Addition of naphthol, polyphenols, aromatic amines, and aminophenol has been proposed for the stabilization of ethyl ether.

1.4.9 Acetone

Additional names: 2-Propanone, Dimethyl ketone; Molecular formula: CH_3COCH_3; Molecular weight: 58.08; Elemental analysis: C 62.04%, H 10.41%, O 27.55%.

Properties: Boiling point 56.53°C; Melting point –94.90°C; Relative density (d_4^{20}) 0.788; Flash point –20°C; Explosive limit 2.5%~12.8% (V/V). Volatile, highly flammable liquid; characteristic odor; pungent, sweetish taste. Miscible with water to form an azeotropic mixture, soluble in organic solvents such as methanol, ethanol, ether, chloroform and pyridine. Repeated exposure to acetone can cause dermatitis. Acetone has better solubility to organic compounds and is a good solvent for refining organic compounds. Acetone can be dried with anhydrous calcium sulfate or anhydrous potassium carbonate, however, should not be dehydrated with sodium metal or phosphorus pentoxide.

1.4.10 Acetonitrile

Additional names: Methyl cyanide, Cyanomethane; Molecular formula: CH_3CN; Molecular weight: 41.05; Elemental analysis: C 58.52%, H 7.37%, N 34.12%.

Properties: Boiling point 81~82°C; Melting point –45.70°C; Relative density (d_4^{20}) 0.787; Flash point 6°C; Explosive limit 3.0%~16.0% (V/V).

A colorless, transparent, volatile and flammable solvent, ether-like odor. Poisonous! It can dissolve a variety of organic, inorganic and gaseous substances with the excellent solubility, miscible with methyl acetate, methyl acetone, ethyl ether, chloroform tetrachloride, carbon tetrachloride and vinyl chloride, and infinitely soluble with water and alcohol. As an important organic intermediate, acetonitrile is used as in the synthesis of vitamin A, cortisone, carbamines and their intermediates, as well as in the production of vitamin B_1 and amino acids. There are also many applications in fabric dyeing, lighting, spice manufacturing and photosensitive material manufacturing. Acetonitrile can be refined by distilling using 5%P_2O_5.

1.4.11 Methylene Chloride

Additional names: DCM, Dichloromethane, Methylene dichloride; Molecular formula: CH_2Cl_2; Molecular weight: 84.93; Elemental analysis: C 14.14%, H 2.37%, Cl 83.49%.

Properties: Boiling point 39.75°C; Melting point −97°C; Relative density (d_4^{20}) 1.326; Refractive index (n_D^{20}) 1.4244; Flash point: free (incombustible); Dielectric constant (ε) 8.9; Explosive limit 12%~19%(V/V). Colorless, volatile and incombustible liquid. Insoluble in water, soluble in ethanol and ether. The low boiling point solvent may cause damage to the nervous system because of its anaesthetic (麻醉的) effect. It is often used as a substitute for flammable petroleum ether and diethyl ether, etc.

Washing with acid, alkali and water in turn, and add anhydrous potassium carbonate for drying, then distillation to get refined products.

1.4.12 Trichloromethane

Additional name: Chloroform, THMS; Molecular formula: $CHCl_3$; Molecular weight: 119.38; Elemental analysis: C 10.06%, H 0.84%, Cl 89.09%.

Properties: Boiling point 61.3°C; Melting point −63.5°C; Relative density (d_4^{20}) 1.48; Refractive index (n_D^{20}) 1.4476. Highly refractive, nonflammable, heavy, very volatile, sweet-tasting liquid; Characteristic odor. Slightly soluble in water; Miscible with alcohol, ether, benzene, petroleum ether, carbon tetrachloride. It can form ternary azeotropic mixture with water and ethanol, boiling point is 55°C, containing 3% water and 4% ethanol. It is oxidized by light energy in the air to produce toxic phosgene. Chloroform should be stored in brown bottles to avoid light, not dry with metal sodium! Otherwise it might explode.

Caution: Potential symptoms of overexposure are dizziness, mental dullness, nausea and disorientation; headache, fatigue; anesthesia; hepatomegaly; direct contact may cause irritation of eyes and skin.

1.4.13 Carbon Tetrachloride

Additional names: Tetrachloromethane, Phenixin; Molecular formula: CCl_4; Molecular weight: 153.82; Elemental analysis: C 7.81%, Cl 92.19%.

Properties: Boiling point 76.8°C; Melting point −22.92°C; Relative density (d_4^{20}) 1.595; Refractive index (n_D^{20}) 1.4603; Dielectric constant (ε) 2.2. Colorless, toxic and volatile liquid, not flammable, with slightly sweet smell of chloroform, dissolve fat and paint, etc. It is strictly limited by the state, used only for non-ozone-depleting substances and special purposes, Not commonly used as an extractant.

It is mildly narcotic (麻醉的) and toxic that cause serious damage to the liver and kidney. Poisoning, such as dizziness, dizziness and listlessness can result from inhalation or contact.

1.4.14 Dichloroethane

Additional names: sym-Dichloroethane, Ethylene dichloride, EDC; Molecular formula:

$ClCH_2CH_2Cl$; Molecular weight: 98.96; Elemental analysis: C 24.27%, H 4.07%, Cl 71.65%.

Properties: Boiling point 83.5°C; Melting point −35.7°C; Relative density (d_4^{20}) 1.26; Refractive index (n_D^{20}) 1.4167; Flash point 13°C; Explosive limit 6.2%~16.0%(V/V).

Colorless, light yellow transparent liquid, pleasant smell; Volatile and flammable with high-density. Miscible with ethanol, chloroform and ether, but insoluble in water. Be narcotic, toxic; May cause skin eczema; Being potentially carcinogenic, its vapor affects vision. Possible solvent alternatives include 1, 3-dioxane and toluene. Used as solvent in wax, fat, rubber and grain insecticide, as well as an intermediate in the synthesis of trichloroethane.

The reagent dichloroethane contains a small amount of acidic impurities, such as water and sodium chloride. In the separating funnel, it is washed with 5% sodium hydroxide aq (or potassium hydroxide aq.), water or concentrated sulfuric acid in turn. The refined product can be obtained by distillation after dehydration of anhydrous calcium chloride or phosphorus pentoxide

1.4.15 Benzene

Additional name: Benzol; Molecular formula: C_6H_6; Molecular weight: 78.11; Elemental analysis: C 92.26%, H 7.74%.

Properties: Boiling point 80.1°C; Melting point 5.5°C; Relative density (d_4^{20}) 0.88; refractive index (n_D^{20}) 1.5011; Flash point −11.1°C. Clear, colorless, volatile, highly refractive and flammable liquid; characteristic odor. Slightly soluble in water, miscible with alcohol, ether, acetone, tetrachloromethane, glacial acetic acid, carbon disulfide and oils. As a non-polar solvent, benzene is often used for extraction, recrystallization and chromatography. Benzene can form an azeotropic mixture with water (bp 69°C, 9% water content) to remove water from the reaction. Benzene is also the basic raw material of petrochemical industry. The yield and technical level of benzene production is one of the symbols of the development of petrochemical industry in China.

Caution: Keeping in well-closed containers in a cool place and away from fire. Potential symptoms of overexposure by inhalation or ingestion are dizziness, headache, vomiting, visual disturbances, staggering gait, hilarity, fatigue, Central Nervous System depression(中枢神经系统抑郁症), and loss of consciousness, respiratory arrest. Chronic exposure has been associated with bone marrow depression and leukemia (白血病). Direct contact may cause irritation of eyes, nose, respiratory system and skin; dermatitis may develop due to defatting action. Aspiration into the lung may lead to chemical pneumonitis.

1.4.16 Toluene

Addition names: Methyl benzene, Phenylmethane; Molecular formula: $CH_3C_6H_5$; Molecular weight: 92.14; Elemental analysis: C 91.25%, H 8.75%.

Properties: Boiling point 110.6°C; Melting point −94.9°C; Relative density (d_4^{20}) 0.866; Refractive index (n_D^{20}) 1.4967; Flash point 4.4°C; Explosive limit 1.2%~7.0%(V/V).

Flammable, highly refractive liquid; Benzene-like odor. Very slightly soluble in water; miscible with alcohol, chloroform, ether, acetone, glacial acetic acid, carbon disulfide. Less toxic than benzene. It can also form an azeotrope with water (bp 85°C), containing approximately 20% water. The amount of water removed is considerable so that toluene is often used to remove the water generated in the reaction. Toluene is widely used as solvent and high octane gasoline additive, as well as an important raw material in organic chemical industry. Be sealed and stored in a cool place.

1.4.17 Pyridine

Additional names: Nitrogen benzene; Molecular formula: C_5H_5N; Molecular weight: 79.10;

Elemental analysis: C 75.92%, H 6.37%, N 17.71%.

Properties: Boiling point 115.2°C; Melting point −41.6°C; Relative density (d_4^{20}) 0.9827; Refractive index (n_D^{20}) 1.5092; Flash point 17°C; Dielectric constant (ε) 12.4; Explosive limit 1.7%~12.4% (V/V). Flammable, colorless/yellowish, weak alkaline liquid; Characteristic disagreeable odor; Sharp taste. Soluble in most organic solvents, such as petroleum ether, alcohol and ether, etc.

Pyridine can be dissolved in water in any proportion, while dissolve most polar and non-polar organic compounds, and even some inorganic salts. Therefore, pyridine is a solvent with wide application. In addition to being a solvent, pyridine can also be used in industry as modifier and dyeing auxiliaries, as well as in synthesis as the starting material for a series of products, including pharmaceuticals, disinfectants, dyes, food seasonings, adhesives, explosives, etc. When used as a catalyst, the dosage should not be excessive, otherwise the product quality will be affected.

Caution: Potential symptoms of overexposure are headache, nervousness, dizziness and insomnia; nausea, anorexia; frequent urination; eye irritation; dermatitis; liver and kidney damage. In some cases of acute poisoning, a nervous breakdown is present. Pyridine poisoning causes fewer deaths.

Dried with solid sodium hydroxide, the aqueous layer is separated out, then add solid sodium hydroxide until the anhydrous layer precipitates. Anhydrous pyridine is obtained by distillation.

1.4.18 *N,N*-Dimethylformamide

Additional name: DMF; Dimethylformamide, *N, N*-dimethyl-formamide; Molecular formula: C_3H_7NO; Molecular weight: 73.09; Elemental analysis: C 49.30%, H 9.65%, N 19.16%, O 21.89%.

Properties: Boiling point 152.8°C; Melting point −61°C; Relative density (d_4^{20}) 0.948; Refractive index (n_D^{20}) 1.4282; Flash point 58°C; Dielectric constant (ε) 36.7; Explosive limit 2.2%~15.2% (V/V). Aprotic polar solvent; Colorless to very slightly yellow liquid; Faint amine odor, hygroscopic. Miscible with water and most organic solvents, such as ethanol, chloroform and ethyl ether, slightly soluble in benzene. Good solubility for many organic and inorganic compounds, and good chemical stability in the absence of base, acid and water. Classified as a low-toxic category. Harmful by inhalation and in contact with skin. Irritating to eyes. May cause harm to the unborn child. Avoid exposure! DMF is not only widely used as a solvent, but also as an important intermediate in organic synthesis. It can be used to produce chlordimeform (杀虫脒) in pesticide field.

DMF often contains impurities, such as water, ethanol, primary amine and secondary amine, and can form $HCON(CH_3)_2 \cdot 2H_2O$ with two molecules of H_2O. To obtain higher purity products, desiccant and distillation may be used together.

1.4.19 Tetrahydrofuran

Additional names: Diethylene oxide, Tetramethylene, THF, Furanidine; Molecular formula: C_4H_8O; Molecular weight: 72.11; Elemental analysis: C 66.63%, H 11.18%, O 22.19%.

Properties: Boiling point 67°C; Melting point −108.5°C; Relative density (d_4^{20}) 0.89; Refractive index (n_D^{20}) 1.4050; Flash point −17.2°C; Dielectric constant (ε) 7.6; Explosive limit 2.0%~11.8% (V/V). Tetrahydrofuran is a heterocyclic organic compound. As one of the strongest polar ethers, tetrahydrofuran is used as a medium polar solvent in chemical reactions and extraction. Colorless transparent liquid, low toxicity, low boiling point, good fluidity, ether-like odor. Miscible with water, alcohol, ketone, benzene, ester, ether and hydrocarbons. Dissolved in water to form an azeotropic mixture containing 5% water (bp 64°C). Highly flammable. May form explosive peroxides. Irritating to eyes and respiratory system. Used as solvent and raw material for organic synthesis.

1.4.20 Dioxane

Additional names: Diethylene dioxide, 1, 4-Diethylene dioxide; Molecular formula: $C_4H_8O_2$; Molecular weight: 88.11; Elemental analysis: C 54.53%, H 9.15%, O 36.32%.

Properties: Boiling point 101.3℃; Melting point 11.8℃; Relative density (d_4^{20}) 1.04; Refractive index (n_D^{20}) 1.4224; Flash point12℃; Dielectric constant (ε) 2.2; Explosive limit 1.97%~25% (V/V). Colorless liquid with a slight fragrance. Soluble in most organic solvents, mixed with water in any proportion, forming an azeotropic mixture containing 18% water (bp 88℃). Highly flammable; May form explosive peroxides; Irritating to eyes and respiratory system; Limited evidence of a carcinogenic effect; Repeated exposure may cause skin dryness or cracking. May cause damage to the liver, kidneys and nervous system. Acute poisoning can lead to death. Mainly used as solvent, emulsifier (乳化剂), detergent and so on.

四、常用有机溶剂

常用有机溶剂的理化常数是学生实验安全和成功进行实验的重要参数，对实验溶剂的性状及基本知识的了解，有助于提高学生对实验过程的理解。

市售有机溶剂通常有分析纯试剂、化学纯试剂、工业试剂等不同规格，可根据实验对溶剂的具体要求来进行选用，一般不需作纯化处理。下列情况需对溶剂进行纯化：实验对溶剂的纯度要求特别高，市售溶剂不能满足要求；溶剂放置时间过长，由于氧化、吸潮、光照等原因可能增加了杂质而不能满足实验要求；溶剂用量过大，为降低成本以较低规格溶剂替代高规格溶剂；溶剂回收再利用。

（一）甲醇

别名：木醇，木精；分子式：CH_3OH；分子量：32.04；元素分析：C 37.48%，H 12.58%，O 49.93%。

性质：沸点 64.96℃，熔点 –97.81℃，相对密度（d_4^{20}）0.792，折射率（n_D^{20}）1.3292，闪点 12℃，爆炸极限 5.5%~44.0%（V/V），介电常数（ε）32.7。甲醇为无色透明的易燃性液体，能与水以任意比例互溶，但不形成恒沸混合物。溶于醇类，乙醚、苯及其他有机溶剂，易挥发、燃烧，有毒，特别是损害视力。吸入后出现头痛、头晕、乏力、眩晕、酒醉感、意识朦胧，呈昏迷状态。视神经及视网膜病变，可有视物模糊、复视等，重者永久性失明。代谢性酸中毒时出现二氧化碳结合力下降、呼吸加速等，甚至死亡。

含水量低于 0.5% 的甲醇，经重蒸馏即可除去水。含水量低于 0.01% 的甲醇，用分馏法或用 4Å 分子筛干燥。

（二）乙醇

别名：酒精，火酒；分子式：C_2H_5OH；分子量：46.07；元素分析：C 52.14%；H 13.13%；O 34.73%。

性质：沸点 78.50℃，熔点 –114.10℃，相对密度（d_4^{20}）0.789，折射率（n_D^{20}）1.3292，闪点 12℃，爆炸极限 3.3%~19.0%（V/V）。在常温、常压下是一种易燃、易挥发的无色透明液体，它的水溶液具有酒香的气味，并略带刺激。有酒的气味和刺激的辛辣滋味，微甘。能与水以任意比例互溶，溶于醇类、乙醚、苯、石油醚等有机溶剂。与水能形成恒沸混合物，恒沸点 78.17℃，含水 4.47%（W/W）。工业酒精含有少量甲醇，医用酒精主要指浓度为 75% 左右的乙醇，也包括医学上使用广泛的其他浓度酒精。

常见的杂质为水、丙酮、甲醛等，市售无水乙醇常含有苯、甲苯等。注意储存和使用保持密闭，冷却，远离火源！乙醇与甲醚是同分异构体。

（三）正丙醇

别名：1–丙醇，n–丙醇，丙醇；分子式：C_3H_7OH；分子量：60.10；元素分析：

C 59.96%，H 13.42%，O 26.62%。

性质：沸点 97.2℃，熔点 –126.5℃，相对密度（d_4^{20}）0.804，闪点 15℃，爆炸极限 2.0%~13.3%（V/V）。正丙醇有似乙醇气味的无色透明液体，溶于水、乙醇和乙醚，可与水形成共沸混合物，恒沸点 87℃，含水量 28.3%（W/W）。主要用来做燃料油的杀菌剂、农药及医药原料、香料原料等。

（四）乙酸乙酯

别名：醋酸乙酯；分子式：$CH_3COOC_2H_5$；分子量：88.11；元素分析：C 54.53%，H 9.15%，O 36.32%。

性质：沸点 77℃，熔点 –84℃，相对密度（d_4^{20}）0.902，闪点 –4℃，折射率（n_D^{20}）1.3719，介电常数 6.0，爆炸极限 2.2%~11.2%（V/V）。无色澄清黏稠状液体，低毒性有刺激性和麻醉性，有甜味，易挥发易燃，对空气敏感，能吸水分（高达 3.3% W/W），使其缓慢水解而呈酸性反应。能与乙醇、丙酮、氯仿和乙醚混溶，溶于水（10% W/W）。能溶解某些金属盐类（如氯化锂、氯化钴、氯化锌、氯化铁等）。其蒸气比空气重，能在较低处扩散到相当远的地方，遇明火会引着回燃。需密闭冷却保存，远离火源。应用于有机化工、香精香料、油漆、医药等行业。

常见杂质为水、乙醇、乙酸。用 5% 碳酸钠溶液洗 1~2 次可洗去酸。用饱和氯化钙溶液可洗去醇，再用无水氧化钙干燥，重蒸即达到精制目的。

（五）正己烷

别名：己烷；分子式：C_6H_{14}；分子量：86.18；元素分析：C 83.63%，H 16.37%。

性质：沸点 68.74℃，熔点 –95.30℃，相对密度（d_4^{20}）0.692，折射率（n_D^{20}）1.388，闪点 30° F (–1.11℃)。无色透明液体具汽油味，易燃，不溶于水，可与乙醚、氯仿混溶，溶于丙酮，在乙醇中的溶解度为 100 份乙醇溶解 50 份正己烷（33℃）。正己烷作为良好的非极性有机溶剂，被广泛使用在化工，机械设备表面清洗去污等环节。但其具有一定的毒性，会通过呼吸道、皮肤等途径进入人体，长期接触可导致人体出现头痛、头晕、乏力、四肢麻木等慢性中毒症状，严重的可导致晕倒、神志丧失甚至死亡。

（六）环己烷

别名：六氢化苯；分子式：C_6H_{12}；分子量：84.16；元素分析：C 85.63%，H 14.37%。

性质：沸点 80.7℃，凝固点 6.5℃，相对密度（d_4^{20}）0.778，折射率（n_D^{20}）1.4266，闪点 –16.5℃，爆炸极限 1.3%~8.3%（V/V）。无色透明液体，有刺激性气味，易挥发易燃，其蒸气比空气重，能在较低处扩散到相当远的地方，遇火源会着火回燃。不溶于水，溶于乙醇、乙醚、苯、丙酮等多数有机溶剂，在甲醇中的溶解度为 100 份甲醇可溶解 57 份环己烷 (25℃)。用作一般溶剂、色谱分析标准物质，可在树脂、涂料、石蜡油类中应用，还可制备环己醇和环己酮等有机物。

（七）石油醚

别名：石油精，石油英；分子式：C_5H_{12}，C_6H_{14}，C_7H_{16}；分子量：主要是戊烷和己烷的混合物；元素分析：大约 C 83.63%，H 16.37%。

性质：沸点30~80℃，沸程可分为30~60℃；60~90℃和90~120℃三种规格，熔点＜–73℃，相对密度（d_4^{20}）0.64~0.66，闪点＜–20° C，爆炸极限1.1%~8.7%（V/V）。

石油醚是无色透明液体，有煤油气味。主要为戊烷和己烷的混合物，非极性有机溶剂。不溶于水，溶于无水乙醇、苯、氯仿、油类等多数有机溶剂。易燃易爆，与氧化剂可强烈反应。其蒸气比空气重，能在较低处扩散到相当远的地方，遇火源会着火回燃。主要用作溶剂和油脂处理。通常用铂重整抽余油或直馏汽油经分馏、加氢或其他方法制得。石油醚不等于汽油，同时，其结构中没有醚键(C–O–C)。该品可引起周围神经炎，对皮肤有强烈刺激性。

（八）乙醚

别名：二乙醚，乙氧基乙烷；分子式：$C_2H_5OC_2H_5$；分子量：74.12；元素分析：C 64.82%，H 13.60%，O 21.59%。

性质：沸点34.6℃，熔点–116.3℃，相对密度（d_4^{20}）0.714，闪点–45℃，爆炸极限1.85%~36.5%（V/V）。无色透明液体，有特殊刺激气味、带甜味，极易挥发易燃易爆、其蒸气重于空气，与空气隔绝时相当稳定，在空气的作用下能氧化成过氧化物、醛和乙酸。微溶于水，可溶于乙醇、氯仿、苯等有机溶剂。蒸馏时不可蒸干，附近严禁有明火；有麻醉性。

杂质多为水、乙醇、过氧化物、醛等。乙醚暴露于空气中或蒸发至干时极易产生过氧化物，可以用5%硫酸亚铁溶液将过氧化物从醚中除去。萘酚、多酚类、芳香胺类、氨基酚类亦可以在乙醚中起到稳定作用。

（九）丙酮

别名：二甲基酮，二甲基甲酮，木酮；分子式：CH_3COCH_3；分子量：58.08；元素分析：C 62.04%，H 10.41%，O 27.55%。

性质：沸点56.53℃，熔点–94.90℃，相对密度（d_4^{20}）0.788，闪点–20℃，爆炸极限2.5%~12.8%（V/V）。无色透明液体，有特殊的辛辣气味。能与水以任意比互溶，形成恒沸混合物，易溶于甲醇、乙醇、乙醚、氯仿和吡啶等有机溶剂。易燃、易挥发、有毒，皮肤长期反复接触可致皮炎。化学性质较活泼。丙酮对有机化合物有较好的溶解度，是精制有机物质的良好溶剂。丙酮可用无水硫酸钙或无水碳酸钾干燥去水，注意丙酮不宜用金属钠或五氧化二磷干燥脱水。

（十）乙腈

别名：甲基氰，氰化甲烷；分子式：CH_3CN；分子量：41.05；元素分析：C 58.51%，H 7.37%，N 34.12%。

性质：沸点81~82℃，熔点–45.70℃，相对密度（d_4^{20}）0.787，闪点6℃，爆炸极限3.0%~16.0%（V/V）。乙腈为无色透明液体，极易挥发且易燃，具有一定毒性，有类似于醚的特殊气味。有优良的溶剂性能，能溶解多种有机、无机和气体物质。可与甲醇、醋酸甲酯、丙酮、乙醚、氯仿、四氯化碳和氯乙烯混溶，与水和醇无限互溶。作为一个重要的有机中间体，乙腈可用于合成维生素A，可的松，碳胺类药物及其中间体的溶剂，还用于制造维生素B_1和氨基酸的活性介质溶剂。在织物染色，照明，香料制造和感光材料制造中也有许多用途。乙腈可与5%P_2O_5一起蒸馏而精制。

（十一）二氯甲烷

别名：甲叉二氯，甲撑氯，亚甲基二氯；分子式：CH_2Cl_2；分子量：84.93；元素分析：C 14.14%，H 2.37%，Cl 83.49%。

性质：沸点 39.75℃，熔点 -97℃，相对密度（d_4^{20}）1.326，折射率（n_D^{20}）1.326，闪点：无（不燃物），介电常数（ε）8.9，爆炸极限 12%~19%（V/V）。无色透明易挥发液体，不燃烧，具有类似醚的刺激性气味。不溶于水，溶于乙醇和乙醚。是不可燃低沸点溶剂，但有麻醉作用，并损害神经系统。常用来代替易燃的石油醚、乙醚等。

依次用酸，碱和水洗涤，加入无水碳酸钾干燥，然后蒸馏，即得精制品。

（十二）三氯甲烷

别名：氯仿；分子式：$CHCl_3$；分子量：119.38；元素分析：C 10.06%，H 0.84%，Cl 89.09%。

性质：沸点 61.3℃，熔点 -63.5℃，相对密度（d_4^{20}）1.48，折射率（n_D^{20}）1.4476。无色透明液体。有特殊气味，味甜。高折光，不燃，质重，易挥发。不溶于水，溶于醇、醚、苯、石油醚和四氯化碳。能与水、乙醇形成三元恒沸混合物，沸点为 55℃，含水 3%，含乙醇 4%。有毒，有麻醉性，长期接触可引起肝脏损伤。在空气中遇光能氧化，产生有毒的光气。注意氯仿应以棕色瓶避光贮存，不得用金属钠干燥，否则可能爆炸！

（十三）四氯化碳

别名：四氯甲烷；分子式：CCl_4；分子量：153.82；元素分析：C 7.81%，Cl 92.19%。

性质：沸点 76.8℃，熔点 -22.92℃，相对密度（d_4^{20}）1.595，折射率（n_D^{20}）1.4603，介电常数（ε）2.2。无色有毒液体，能溶解脂肪、油漆等多种物质，易挥发液体，不易燃，具氯仿的微甜气味。与水互不相溶，可与乙醇、乙醚、氯仿及石油醚等混溶。四氯化碳的用途被国家严格限制，仅限用于非消耗臭氧层物质原料用途和特殊用途，作为萃取剂并不常用。

有轻微麻醉性，有毒，对肝和肾能引起严重的损害，吸入或接触均可导致中毒，慢性中毒症状为头晕、眩晕、倦怠无力等。

（十四）1, 2- 二氯乙烷

别名：邻 - 二氯乙烷，乙撑二氯；分子式：$ClCH_2CH_2Cl$；分子量：98.96；元素分析：C 24.27%，H 4.07%，Cl 71.65%。

性质：沸点 83.5℃，熔点 -35.7℃，相对密度（d_4^{20}）1.26，折射率（n_D^{20}）1.4167，闪点 13℃，爆炸极限 6.2%~16.0%（V/V）。无色或浅黄色透明液体，有愉快的气味，易挥发易燃烧，密度大。能与乙醇、氯仿和乙醚混溶，难溶于水。有麻醉性，有毒，能引起皮肤湿疹，其蒸气影响视力，具潜在致癌性。可能的溶剂替代品包括 1, 3- 二氧杂环己烷和甲苯。用作溶剂，还用于蜡、脂肪、橡胶等的溶剂及谷物杀虫剂。

试剂二氯乙烷中含有少量酸性杂质、水分、氯化钠等，在分液漏斗中依次用 5% 氢氧化钠溶液 (或氢氧化钾溶液)、水洗涤或浓硫酸洗涤，无水氯化钙或五氧化二磷脱水后蒸馏，可得精制品。

（十五）苯

别名：安息油；分子式：C_6H_6；分子量：78.11；元素分析：C 92.26%，H 7.74%。

性质：沸点 80.1℃，熔点 5.5℃，相对密度（d_4^{20}）0.88，折射率（n_D^{20}）1.5011，闪点 −11.1℃。

无色透明液体，有芳香气味。具强折光性。易挥发易燃。能与乙醇、乙醚、丙酮、四氯化碳、二硫化碳、冰乙酸和油类任意混溶，微溶于水。苯是非极性溶剂，常用来提取、重结晶和层析有机化合物。苯和水能形成共沸混合物（b. p. 69℃，含水量为 9%），故常利用苯的这种性质来除去反应中生成的水。苯也是石油化工的基本原料，苯的产量和生产的技术水平是一个国家石油化工发展水平的标志之一。

苯的毒性高，是一种致癌物质，长期接触会引起慢性中毒，主要表现为破坏人体造血功能。通风低温干燥密闭储存，与氧化剂分开存放。

（十六）甲苯

别名：甲基苯，苯基甲烷；分子式：$CH_3C_6H_5$；分子量：92.14；元素分析：C 91.25%，H 8.75%。

性质：沸点 110.6℃，熔点 −94.9℃，相对密度（d_4^{20}）0.866，折射率（n_D^{20}）1.4967，闪点 4.4℃，爆炸极限 1.2%~7.0%（V/V）。无色透明液体，有类似苯的芳香气味，强折光性。毒性比苯小，属低毒类。能与乙醇、乙醚、丙酮、氯仿、二硫化碳和冰乙酸混溶，极微溶于水，也能和水形成共沸混合物（b. p. 85℃），约含 20% 的水，因此甲苯的除水量相当大，加上它本身又是一个较好的溶剂，因此在实验中经常用甲苯除去反应中生成的水。甲苯大量用作溶剂和高辛烷值汽油添加剂，也是有机化工的重要原料。注意密封阴凉保存。

（十七）吡啶

别名：氮（杂）苯；分子式：C_5H_5N；分子量：79.10；元素分析：C 75.92%，H 6.37%，N 17.71%。

性质：沸点 115.2℃，熔点 −41.6℃，相对密度（d_4^{20}）0.9827，闪点 17℃，折射率（n_D^{20}）1.5092，介电常数（ε）12.4，爆炸极限 1.7%~12.4%（V/V）。吡啶是含有一个氮杂原子的六元杂环化合物，为无色或微黄色液体，显弱碱性，易燃，有恶臭和强刺激性。溶于石油醚、醇和醚等多数有机溶剂。吡啶与水能以任何比例互溶，同时又能溶解大多数极性及非极性的有机化合物，甚至可以溶解某些无机盐类，所以吡啶是一个有广泛应用价值的溶剂。除作溶剂外，吡啶在工业上还可用作改性剂、助染剂，以及合成一系列产品（包括药品、消毒剂、染料、食品调味料、黏合剂、炸药等）的起始物，做催化剂时用量不可过多，否则影响产品质量。

对皮肤有刺激性，吸入蒸汽可出现头晕、恶心及肝脏损坏，大量吸入能麻痹中枢神经。一些急性中毒事件中表现为精神崩溃。吡啶中毒引起死亡的事件比较少。可用固体氢氧化钠干燥，分离析出的水层后，再加固体氢氧化钠至无水层析出，然后蒸馏即得无水吡啶。

（十八）*N, N*– 二甲基甲酰胺

别名：*N*– 甲酰二甲胺，二甲基甲酰胺，DMF；分子式：C_3H_7NO；分子量：73.09；元素分析：C 49.30%，H 9.65%，N 19.16%，O 21.89%。

性质：沸点 152.8℃，熔点 −61℃，相对密度（d_4^{20}）0.948，折射率（n_D^{20}）1.4282，闪点 58℃，介电常数（ε）36.7，爆炸极限 2.2%~15.2%（V/V）。非质子型极性溶剂，无色透明或淡黄

色液体，有鱼腥味，有吸湿性。能与水、乙醇、氯仿和乙醚等多数有机溶剂混溶，微溶于苯。对多种有机化合物和无机化合物均有良好的溶解能力，在无碱、酸、水存在下，具有良好的化学稳定性。属低毒类，吸入及皮肤接触有害，刺激眼睛，可能对胎儿造成伤害。避免接触！在有机反应中，二甲基甲酰胺不但广泛用作反应的溶剂，也是有机合成的重要中间体，在农药中还可用来生产杀虫脒。

N, N-二甲基甲酰胺常含有水、乙醇、伯胺、仲胺等杂质，并能与两分子水形成[$HCON(CH_3)_2 \cdot 2H_2O$]水合物。要得到高纯度产品，可使用干燥剂与蒸馏并用的方法。

（十九）四氢呋喃

别名：氧杂环戊烷，1, 4-环氧丁烷；分子式：C_4H_8O；分子量：72.11；元素分析：C 66.63%，H 11.18%，O 22.19%。

性质：沸点67℃，熔点-108.5℃，相对密度（d_4^{20}）0.89，闪点-17.2℃，折射率（n_D^{20}）1.4050，介电常数（ε）7.6，爆炸极限2.0%~11.8%（V/V）。四氢呋喃是一类杂环有机化合物。它是最强的极性醚类之一，在化学反应和萃取时常作为中等极性的溶剂。无色透明液体，具有低毒、低沸点、流动性好等特点，有类似乙醚的气味。与水、醇、酮、苯、酯、醚、烃类混溶，溶于水并能形成含水5%、沸点为64℃的恒沸混合物。高度易燃，可能生成爆炸性过氧化物，刺激眼睛和呼吸系统。用作溶剂、有机合成的原料。

（二十）二氧六环

别名：二恶烷，1, 4-二氧己环，1, 4-二氧杂环己烷；分子式：$C_4H_8O_2$；分子量：88.11；元素分析：C 54.53%，H 9.15%，O 36.32%。

性质：沸点101.3℃，熔点11.8℃，相对密度（d_4^{20}）1.04，折射率（n_D^{20}）1.4224，闪点12℃，介电常数（ε）2.2，爆炸极限1.97%~25%（V/V）。无色液体，稍有香味。溶于多数有机溶剂，能与水以任意比例互溶，能形成含水18%，沸点88℃的恒沸混合物。高度易燃，可能生成爆炸性过氧化物。属微毒类，刺激眼睛和呼吸系统，少数报道有致癌后果，长期接触可能引起皮肤干裂，可能对肝、肾和神经系统造成损害，急性中毒时可能导致死亡。主要用作溶剂、乳化剂、去垢剂等。

1.5 Experimental Preview, Experimental Record and Laboratory Report

1. Experimental preview record Make good preparations before you come to the lab by reviewing the theory and seeking information about the chemicals involved. If you prepare well, you will save time and effort. In general, the experimental record includes the following parts:

(1) Experimental purpose, principle, instrument and general statement of the process to be done.

(2) Note any special observations or precautions required.

(3) Jot down any chemical or biological hazards it presents, how to deal with it.

(4) Consult main physical contents of the main reagents and the raw materials in this experiment

2. Experimental report Usually, the experimental report should cover the following parts.

(1) Experiment title.

(2) Experimental skills; main techniques introduced by the experiment.

(3) Principle of the experiment; chemical reaction; balanced chemical equations or reaction mechanism.

(4) Experimental apparatus and reagent.

(5) Experimental procedure process and experimental observations recorded.

(6) Result and discussion analysis of your data and brief comments on sources of error.

(7) Results.

(8) Questions.

五、实验预习、实验记录和实验报告

1. 实验预习记录 在进入实验室之前必须通过预习有关化学实验理论知识及相关信息等，做好充分的准备。如果预习充分，实验将会省时高效。通常实验预习、记录包括如下几个部分。

（1）实验目的、原理、仪器和实验过程。

（2）实验注意事项。

（3）实验中存在的危险，如何处理。

（4）查阅本实验的主要试剂及原料主要物理常数。

2. 实验报告 通常实验报告应包括下列几个部分。

（1）实验题目。

（2）实验技能；实验的主要操作或技术。

（3）实验原理；化学反应和反应机理等。

（4）实验仪器和主要试剂。

（5）实验过程 实验步骤和现象记录。

（6）结果和讨论 数据处理及误差分析。

（7）结论。

（8）思考题。

Part Ⅱ 第二部分

Basic Techniques

基本操作

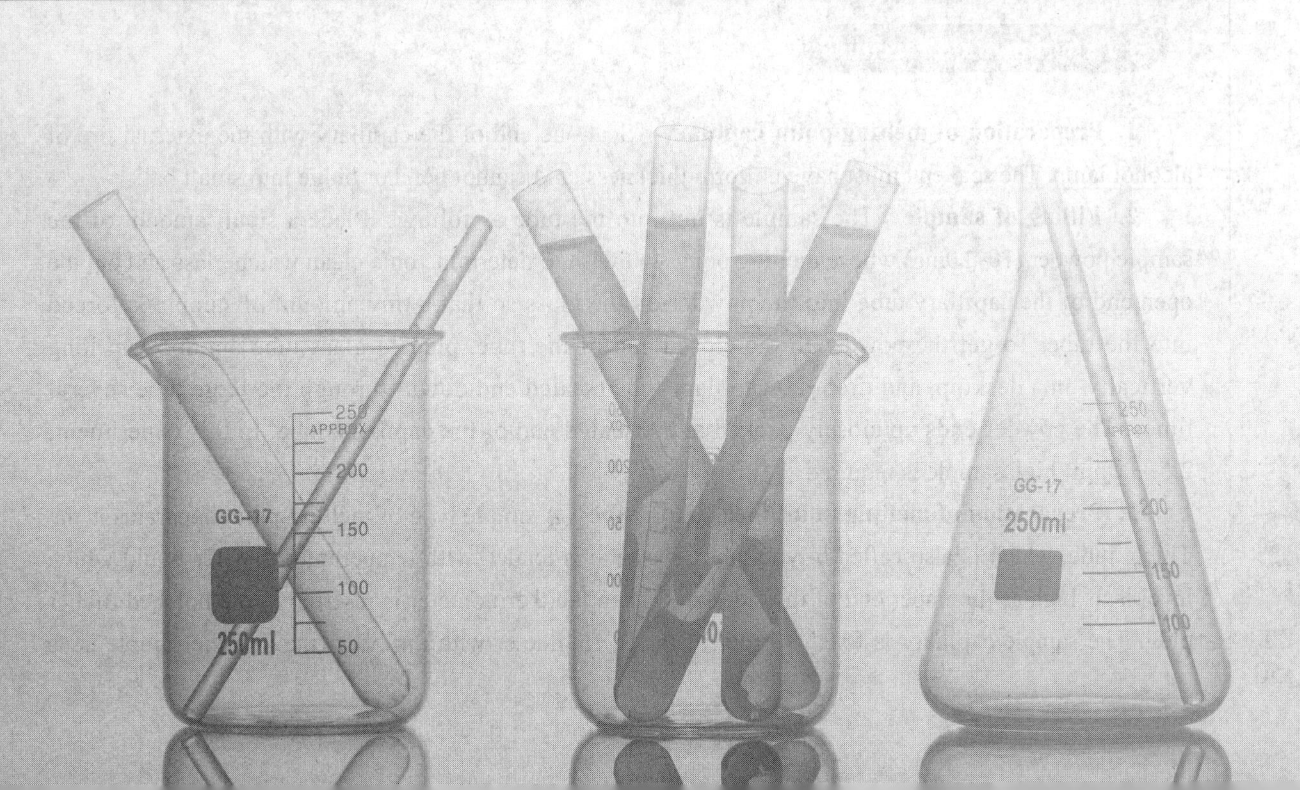

1 Melting Point Determination

Experimental Purpose

1. Master the method of melting point (熔点) determination of solid organic compounds.
2. Understand the principle of testing the purity of organic substance by melting point determination.

Experimental Principle

When the vapor pressure of the solid is equal to that of the liquid, the temperature is the melting point. Most pure organic compounds have a certain melting point, and the melting point distance (i. e. the starting melting temperature to the full melting temperature) is also very short, with a difference of only 0.5~1.0℃. However, if there is a small amount of impurities, the melting point distance will increase and the melting point will decrease. Therefore, the determination of melting point is often used to identify substances and qualitatively test the purity of substances.

Instruments and Reagents

1. Instruments Thiele tube (提勒管), thermometer (温度计), alcohol lamp, melting point capillary, rubber plug, rubber ring.

2. Reagents liquid paraffin (石蜡), benzoic acid (苯甲酸), acetanilide (乙酰苯胺).

Experimental Procedure

1. Preparation of melting point capillary Seal one end of the capillary with the external fire of alcohol lamp. The seal end must have uniform thickness, and cannot bend or bulge into small balls.

2. Filling of sample The sample is put into the tube as follows. Place a small amount of the sample powder (10~20mg) whose melting point you wish to determine on a clean watch glass and tap the open end of the capillary tube into the powder on the glass so that a tiny amount of sample is forced into the tube. To get the powder to the closed end of the tube, place a glass tube about 50cm long vertically on a desktop, and drop the capillary tube (sealed end down) through the large tube several times. The powder ends up closely packed at the sealed end of the capillary tube. In this experiment, 2.5~3.5mm high sample is required.

3. Preparation of melting point measuring tube A simple type of melting-point apparatus is the Thiele tube, which is also called b-type tube. The tube is loaded with temperature transfer liquid which level is as high as the upper end of the side tube. Then the thermometer is inserted into a notched rubber plug. The sample capillary is fixed in front of the thermometer with a rubber ring and the sample is as

high as the middle of the mercury sphere (rubber ring cannot touch the temperature transfer liquid). Put the sample in the middle of the left side of the Thiele tube. The apparatus is shown in Figure 2-1.

4. Determination Heat the triangle tip of the tube. At the beginning, the temperature can rise by 5~6℃ per minute. When the temperature is heated to 10~20℃ below the expected melting point, use small fire and continue heating the Thiele tube at the rate of 1~2℃ per minute in order to determine the melting point. At the same time, observe the change in temperature and sample. When the sample in the capillary tube begins to hair, round, concave or change shape, have droplets, the temperature is initial melting point. When the sample has melted completely, the temperature is final melting point. The melting temperature from the beginning to the end is the melting point, and the difference between the two is the melting point distance.

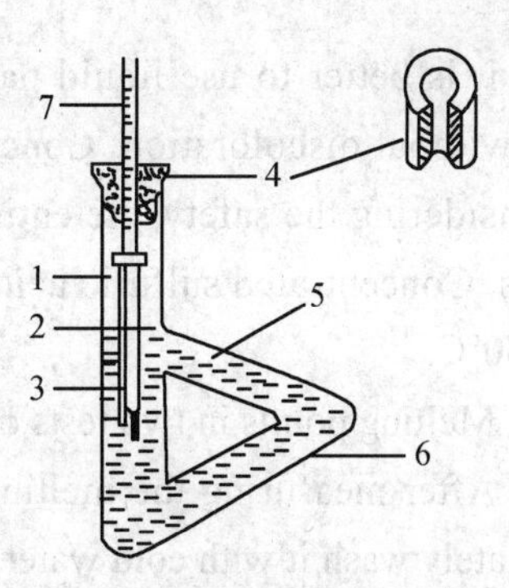

Figure 2-1 Melting Point Determination Apparatus

图 2-1 熔点测定装置图

1. high temperature liquid level; 2. room temperature liquid level; 3. melting point capillary; 4. notch rubber plug; 5. solution; 6. alcohol lamp; 7. thermometer

For the samples that are easy to decompose, you can preheat the hot bath to a temperature which are about 20℃ below the melting point before installing sample capillary, and then use a small fire for heating to measure melting point.When the hot bath temperature drops by 30℃ or so, replace the melting point tube, and then determine the melting point again.

5. Experimental samples ① acetanilide; ② benzoic acid; ③ mixed samples of acetanilide and benzoic acid; ④ benzoicacid and so on. Each sample is roughly measured once and accurately measured twice, and the experimental datas are recorded.

Samples	Acetanilide		Benzoic Acid	
	Exp. 1	Exp. 2	Exp. 1	Exp. 2
Melting-Point Range/℃				

Preview Guide

Preview Requirements

1. Understand the principle and significance of melting point measurement.
2. Learn the operation method of melting point measurement.
3. Explain the relationship between the melting range and the purity of solids.

Notes

1. Choose different hot bath according to the different temperature measured. Frequently used liquids are concentrated sulfuric acid, glycerin, liquid paraffin and so on. If the temperature is below

140°C, it is better to use liquid paraffin or glycerin. The medicinal liquid paraffin can be heated to 220°C without discoloration. Concentrated sulfuric acid can be used for higher heating temperature. But considering the safety, the entire operation process should be careful, and the operator must wear goggles. Concentrated sulfuric acid containing potassium sulfate can be used for heating temperature over 250°C.

2. Melting points in two tests of the pure substance cannot exceed ± 1°C.

3. After measuring the melting point, do not take out the thermometer immediately, and do not immediately wash it with cold water to avoid the thermometer burst.

Experimental Explanation

1. When installing melting point measuring instrument, paper can be spread on the experimental table to avoid the liquid paraffin dripping on the experimental table which is difficult to clean.

2. During the test, the rubber ring should always be kept higher than the liquid paraffin and should not be immersed in it.

3. When the temperature is near the melting point, pay attention to the rate of temperature rise. Otherwise the melting process cannot be seen, the experiment fails. Usually, the alcohol lamp must be removed at the beginning of melting, and the remaining solid can be melted by the inertia of heating up.

Questions

1. What is the effect of heating rate on the determination result?

2. For the same sample, does the melting point tube need to be replaced after the first measurement?

3. Samples A and B are both white powders, and the melting points are all measured at 150°C. Are samples A and B the same substance?

实验一 熔点测定

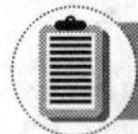

实验目的

1. 掌握测定固体有机物熔点的方法。
2. 理解用熔点测定法来检验有机物纯度的原理。

实验原理

当物质的固态与液态蒸气压相等时的温度，即为熔点。纯粹的有机化合物一般都有一定熔点，且熔点距（即开始熔化到完全熔化的温度）也很短，只相差 0.5~1.0℃。但如有少量杂质存在，物质的熔点距增大，并使熔点降低。因此熔点的测定常可用来识别物质及定性地检验物质的纯度。

视频

仪器和试剂

1. 仪器 提勒管、温度计、酒精灯、熔点毛细管、橡皮塞、橡皮圈。

2. 试剂 液状石蜡、苯甲酸、乙酰苯胺。

实验步骤

1. 熔点毛细管的制备 在酒精灯火焰的边缘将毛细管的一端封头，封口处不能弯曲、鼓成小球，而且厚薄要均匀。

2. 样品的填装 取少量待测样品（10~20mg）粉末于一干净表面皿上，堆成小堆。将毛细管开口一端插入其中数次，这时就有少许样品挤入毛细管中。将一根长约 50cm 的玻璃管直立地放在桌面上，将装有样品的毛细管（闭口端在下）由玻璃管上端自由落下，反复几次，样品就能紧密地填装在毛细管底部，高 2.5~3.5mm。

3. 熔点测定管的准备 熔点测定管又称提勒管或 b 形管，装入传温液，其量与侧管口上端相平，将温度计插入有缺口的橡皮塞中，装有样品的毛细管用一橡皮圈固定在温度计正前方，高低在水银球中部（橡皮圈不能触及传温液）。将温度计插入在熔点测定管的上下两侧管的中部。所得装置如图 2-1 所示。

4. 加热测定 在三角口的尖端加热，刚开始时，温度每分钟可上升 5~6℃，加热到与所预期的熔点相差 10~20℃ 时，改用小火，继续加热，使每分钟上升 1~2℃。同时观察温度计示数与样品变化的情况。当毛细管内样品开始发毛，发圆，发凹或形状改变出现有液滴时，这时的温度为始熔点，至全部透明时为终熔点。始熔至终熔的温度即为熔点，两者的差数为熔点距。

一般容易分解的样品，可把热浴预热至熔点前 20℃ 左右时再装上样品的毛细管，并改用小火加热测熔点。等热浴温度下降 30℃ 左右，重新取熔点管装样，再加热和测定第二次。

5. 测定样品 ①乙酰苯胺；②苯甲酸；③乙酰苯胺；④苯甲酸等样品。

每一样品粗测一次，精密测定两次，记录实验数据。

样品	乙酰苯胺		苯甲酸	
	第一次	第二次	第一次	第二次
熔程 /℃				

预 习 指 导

预习要求

1. 理解熔点测定的原理及意义。
2. 学习熔点测定操作方法。
3. 解释熔程和固体纯度的关系。

注意事项

1. 根据测定温度不同，选择不同热浴，通常有浓硫酸、甘油、液状石蜡等。如果温度在140℃以下，最好用液状石蜡或甘油，药用液状石蜡可加热到220℃仍不变色；需要加热温度较高时，可用浓硫酸，但用浓硫酸不太安全，整个操作过程都应小心，并戴护目镜；温度超过250℃时，可在浓硫酸中加入硫酸钾。

2. 纯净物，两次测定误差不能超过 ±1℃。

3. 在测完熔点后，勿立即取出温度计，更不能立即用冷水冲洗，以免温度计爆裂。

实验说明

1. 在安装熔点测定仪器时，为避免液状石蜡滴在实验台上不好清理，可在实验台上铺张卫生纸。

2. 在测定过程中，橡胶圈应始终保持高于液状石蜡，而不可浸入其中。

3. 接近熔点时，注意升温速度，否则看不到熔化的过程，实验失败。通常，始熔必须撤火，靠升温的惯性即可将剩余固体熔化。

思考题

1. 测定时加热速度对测定结果有何影响？
2. 同一样品，第一次测完后，是否需要重新换毛细管及样品？
3. 样品A和B均为白色粉末，分别测得熔点各为150℃，A和B是同一种物质吗？

2 Boiling Point Determination

Experimental Purpose

1. Understand the principles of boiling point (沸点) determination by micro method.
2. Master the operation of boiling point determination by micro method.

Experimental Principle

Boiling point of liquid: When the liquid substance is heated, the saturated vapor pressure of the liquid increases. When the vapor pressure increases to the same level as the external pressure, the liquid begins to boil. The temperature at this time is called the boiling point of the liquid. Most pure liquid compounds have a certain boiling point, and the boiling point distance (i. e. the temperature from initial boiling to full boiling) is also very short, with a difference of 0.5~1.0℃. But if there is a small amount of impurities, the boiling point distance of the substance will increase. The boiling point of liquid can be determined by constant method (distillation method) and micro method.

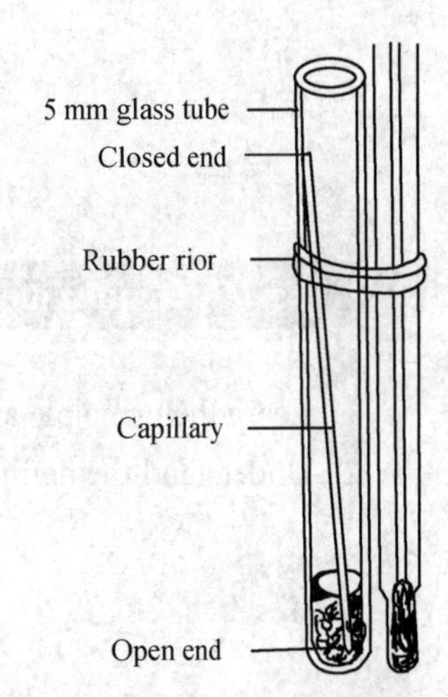

Figure 2-2 Boiling Point Determination Apparatus by Micro Method

图 2-2 微量沸点测定装置

The boiling point tube will be used in micro method. As shown in Figure 2-2, when the liquid is heated, the gas in the capillary tube will continuously escape due to thermal expansion. When the capillary is mainly occupied by the liquid vapor, the liquid can be cooled. Then the vapor pressure of the liquid will drop. Once the vapor pressure reduced to slightly smaller than external pressure, the liquid will enter the capillary tube. At this time, the temperature is the boiling point of the liquid. Only a small amount of liquid is required to obtain accurate determination results by micro method. constant method will be introduced in Experiment4.

Instruments and Reagents

1. Instruments Thiele tube, boiling tube, capillary, thermometer, rubber plug, rubber ring, alcohol lamp.

2. Reagent liquid paraffin, ethyl acetate (乙酸乙酯).

Experimental Procedure

The thermometer is attached to a boiling point tube with a rubber band or copper wire, and the liquid to be measured is placed in the boiling point tube with a height of about 6~7mm. A capillary tube with a length of about 8cm and one end sealed is inserted into the boiling point tube in reverse. Then insert the thermometer into the middle part of the Thiele tube. Considering the thermal expansion of liquid, the rubber ring should not be immersed in the hot bath and should be away from the liquid surface of the hot bath. Typical apparatus is shown in Figure 2-2.

Heat the triangle part of the Thiele tube slowly until a series of small bubbles emerge in the capillary tube, and then stop heating and let the hot bath to cool naturally. When the liquid no longer bubbles or the liquid begins to enter the capillary tube, it is the boiling point of the sample.

When the liquid stops bubbling or begins to enter the capillary tube, it is the boiling point of the sample. After the hot bath temperature drops about 30°C, replace the boiling tube and capillary, and then determine the boiling point again.

Ethyl Acetate	Exp. 1	Exp. 2	Exp. 3	Average
Boiling Point /°C				

Preview Guide

Preview Requirements

1. Learn the principle and operation of measuring boiling point by micro method.
2. Understand the definition of boiling point.

Notes

1. For the low boiling point liquids, when the bubbles in the capillary are not too much, the heating can be stopped and the temperature can be lowered.

2. After measuring the boiling point, do not take out the thermometer immediately, and do not wash with cold water immediately to avoid the thermometer burst.

3. For the unknown liquid samples, the boiling point should be measured roughly. After knowing the approximate boiling point range, the boiling point should be measured accurately.

Experimental Explanation

1. During the heating process, the rubber ring should always be higher than the liquid paraffin level and should not be immersed into the liquid paraffin.

2. The heating speed should not be too fast and the liquid under test should not be too little to prevent the liquid from all gasification.

3. The air inside capillary should be as clean as possible. Before formal determination, numerous bubbles emerging in capillary tube should be emitted to bring out the air.

Questions

If the air is not removed thoroughly after heating, how does it affect the determination result of the boiling point?

实验二 沸点测定

实验目的

1. 理解微量法测定沸点的原理。
2. 掌握微量法测定沸点的操作。

实验原理

视频

液体沸点：当液态物质受热时，液体的饱和蒸气压增大，待蒸气压增至和外压相等时，液体开始沸腾，此时的温度称为液体的沸点。纯粹的液态化合物大多有一定沸点，同时沸程（即开始沸腾到完全沸腾的温度）也很短，只相差 0.5~1.0℃。但如有少量杂质存在，物质沸程就增宽。液体沸点测定可用常量法（蒸馏法）和微量法。

微量法测定沸点要用沸点管，如图 2-2 所示，加热液体时，毛细管内气体受热膨胀会不断逸出，液体沸腾后毛细管内主要由液体蒸气占据。这时，冷却液体，液体蒸气压将下降，当降至微小于外界压力时，液体将进入毛细管内，此时的温度即为液体的沸点。微量法仅需少量液体即可得到准确的测定结果。常量法在实验四中介绍。

仪器设备和试剂

1. 仪器 提勒管、沸点管、毛细管、温度计、橡皮塞、橡皮圈、酒精灯。

2. 试剂 液状石蜡、乙酸乙酯。

实验步骤

温度计上用橡皮圈或铜丝附一沸点管，内放待测液高 6~7 毫米，将长约 8cm 一端封口的毛细管倒插在沸点管中。然后将温度计插入在提勒管的中部。橡皮圈不应浸入热浴，考虑到液体加热膨胀，橡皮圈应离热浴液体面远一些。装置图如图 2-2。

慢慢加热提勒管三角处，至毛细管中有一连串小气泡冒出，此时即刻停止加热，让热浴自行冷却。当液体不再冒气泡或者液体开始进入毛细管时的温度，即为样品的沸点。

待热浴温度下降 30℃ 左右，更换沸点管和毛细管后，再测第二次。

乙酸乙酯	第一次	第二次	第三次	平均值
沸点 /℃				

预习指导

预习要求

1. 学习微量法测定沸点的原理和方法。
2. 理解沸点的定义。

注意事项

1. 对于低沸点液体，当毛细管中气泡冒出不太多时，即可停止加热，使温度下降。
2. 测完沸点后，不要马上取出温度计或用冷水冲，以防温度计破裂。
3. 对于未知样品要粗测沸点，知道大致沸点范围后，再精测。

实验说明

1. 升温过程中，橡胶圈始终应高于液状石蜡液面，不能浸入液状石蜡中。
2. 加热不能过快，被测液体不宜太少，以防液体全部气化。
3. 毛细管内的空气要尽量赶干净，正式测定前，让毛细管里有大量气泡冒出，以此带出空气。

思考题

若加热后毛细管内空气赶除不净，对微量法测定沸点结果有何影响？

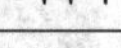

3 Recrystallization

Experimental Purpose

1. Master the principles and methods of recrystallization (重结晶).
2. Master basic operations of recrystallization, such as filtration (过滤), reflux (回流), etc.

Experimental Principle

Any product obtained from reactions is mixed with impurities, such as inactive raw materials and by-products. In order to obtain relatively pure substances, the products must be refined. Organic compounds generally exist in two forms, solid or liquid, and their refining methods are also different. Recrystallization is the most common method to purify solids, while sublimation is often used to purify materials that are easy to sublimate.

The general process of recrystallization is as follows. Dissolve the solid substance in a suitable solvent. The solution is filtered (decolorized if necessary) to remove impurities, and then the solution is concentrated, cooled or otherwise treated to precipitate the purer crystals. The filtrate obtained after filtration is mother liquor, and the crystals obtained are purified substances. Sometimes this operation needs to be repeated to obtain more pure substances.

1. Selecting the appropriate solvent It is a key problem in recrystallization. The selected solvent cannot react with the recrystallized substance. The solubility of the recrystallized substance and impurities in the solvent is quite different. Generally, the solubility of the recrystallized substance varies significantly with the temperature, which is good at higher temperature while poor at lower temperature. And the solvent is easy to volatilize and separate from recrystallized substances. In addition, the toxicity, flammability and price of solvents should also be properly considered.

The solvents commonly used in recrystallization are water, ethanol, methanol, diethyl ether, petroleum ether, glacial acetic acid and benzene, etc. In order to select suitable solvents, the composition and structure of the soluble substances must be considered. In addition to checking chemical manuals, sometimes we should do solubility test.

Solubility test method: Take 0.1g of solid for recrystallization. Put it into a small test tube, add a certain kind of solvent dropwise, and then shake the test tube constantly. When the volume of solvent added is close to 1ml, heat the mixture to boil (note the flammability of the solvent). If the solid is readily soluble in the boiling solvent, the solvent is not suitable.

If the sample is insoluble in boiling solvent, the solvent can be added step by step, the volume is about 0.5ml each time, and continue heating to boiling.

If the substance is still insoluble when the volume of solvent added reaches 3ml, the solvent is not applicable. If the substance dissolve in the hot solvent within 3ml, the mixture will be cooled to observe

whether there is crystallization. If necessary, a glass rod can be used to help crystallization by rubbing inner wall of test tube.

If there is no crystal precipitation within a few minutes after immersion in cooling liquid and rubbing the inner wall of the test tube with a glass rod, this means that the solvent is not suitable for use alone in recrystallization.

According to the above method, different solvents can be tested one by one in many clean test tubes. The most suitable solvent is that in which most crystals can be obtained after cooling. Then it can be added to the sample in proper proportion for crystallization.

When a suitable solvent is not selected, the mixed solvent is usually used. This is a pair of mutually soluble solvents. The compound to be crystallized is soluble in one of the solvents, but insoluble in the other.

Commonly used mixed solvents are water-ethanol, water-acetic acid, water-acetone, water-pyridine, petroleum ether-benzene, petroleum ether-diethyl ether, petroleum ether-acetone, diethyl ether-ethanol, etc.

2. Heating Recrystallization often requires preparation of saturated solution of sample. If water is used as solvent, it can be heated in an open container. If organic solvents are used, reflux heating method (as shown in Figure 2-3) is required to prevent the loss of solvent and the risk of combustion and explosion.

3. Filtration In recrystallization, the commonly used filters are hot filtration and suction filtration.

(1) Hot filtration. In order to prevent the main solute precipitation in the filtration process, it is necessary to filter while hot. A hot water funnel is usually used. After it is fixed and installed properly, the water in the jacket is pre-heated and then filtered, as shown in Figure 2-4. If the amount of solution is small, a short and thick neck glass funnel can be chosen to put in oven for preheating, and then the solution is filtered while hot after taken out. In order to speed up the filtration process, folding filter paper is often used. The folding method is shown in Figure 2-5. Take a circular filter paper and fold it in half, then fold it in half continuously between two creases, and finally open it to obtain a uniform 16-equal sector, as shown in Figure 2-5.

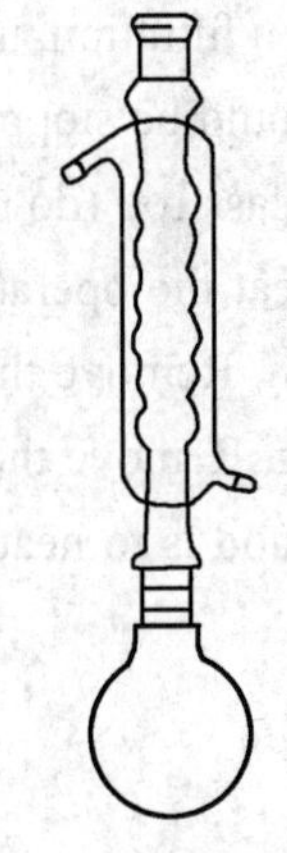

Figure 2-3 Reflux Apparatus
图 2-3 回流装置图

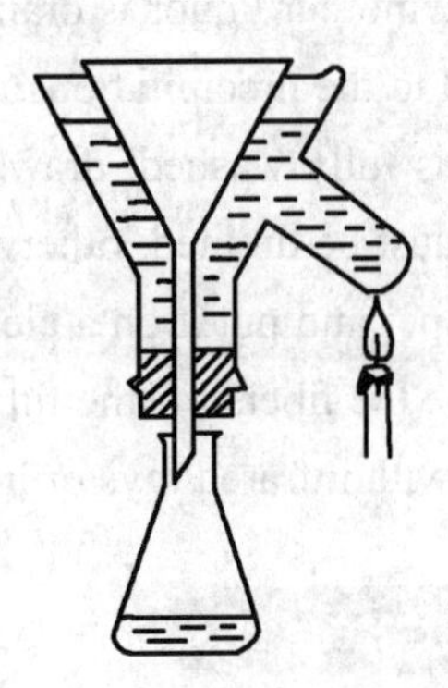

Figure 2-4 Hot Filtration Apparatus
图 2-4 热过滤装置

(2) Suction filtration. It can accelerate the speed of filtration, and separate crystal from mother liquor as much as possible. Buchner funnel and suction filter bottle are commonly used (using glass nails when crystals less than 0.5g). The apparatus is shown in Figure 2-6. Vacuum pumps are often used for suction power.

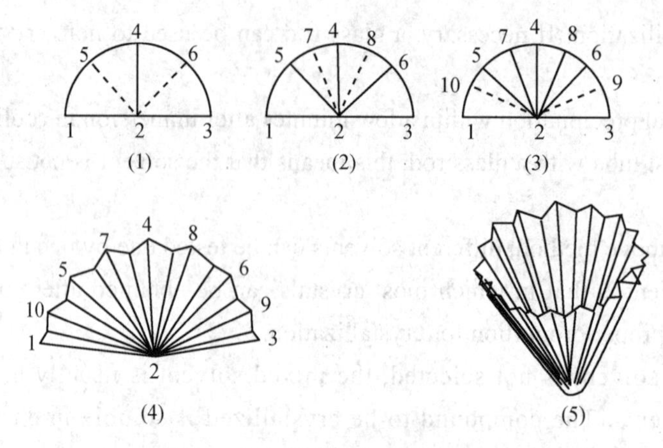

Figure 2-5 Foldable Filter Paper

图 2-5 折叠式滤纸

4. Crystallization There are two main methods. One is to reduce the volume (concentrate) of solvents, which can be evaporated naturally or heating evaporation. The second is to lower the temperature of the solution. If the selected solvent not only can remove impurities, but also has a significant difference on the solubility of thc samplc through the change of temperature (for example, the sample dissolves in hot solvents but hardly dissolves in cold solvents). Thus, the sample can be dissolved in the minimum amount of boiling solvents, and then be filtered immediately to remove insoluble impurities and release more pure crystals after cooling.

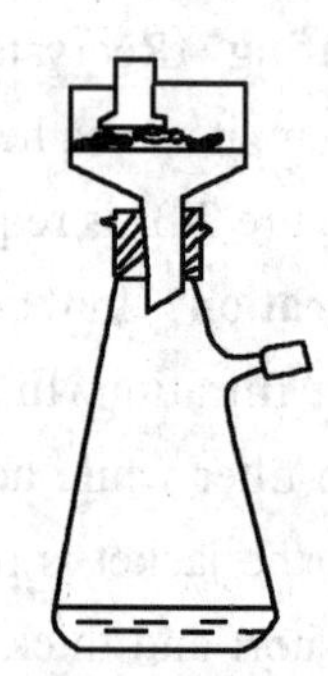

Figure 2-6 Suction Filtration Apparatus

图 2-6 抽滤装置

5. Crystal washing and drying After crystal filtration, in order to separate the crystal from the mother liquor as far as possible, it should be fully washed during suction filtration, usually on the Buchner funnel. That is, after the mother liquor is drained, the air extraction should be stopped, a small amount of washing liquid is added to the precipitation, and then agitated with a glass rod (do not break filter paper). After the precipitation is fully washed, draw away the detergent, repeat the operation 2~3 times. Every time the precipitation must be drained, otherwise it is not easy to clean. Remove the precipitate from the funnel with the filter paper and put it on a clean watch glass for drying. Remove the wet filter paper with a tweezer (do not scrape the fibers off the filter paper). The usual method is to heat it in a water bath, an electric oven, or to dry with infrared rays or in a dryer.

Instruments and Reagents

1. Instruments 50ml conical bottle, 250ml beaker, triangular funnel, filter paper, Büchner funnel, suction filter bottle, glass rod, watch glass, water pump, drying box, iron shelf, iron ring, 50ml round-bottom flask, spherical condenser tube.

2. Reagents crude acetanilide (乙酰苯胺), activated carbon, 33% ethanol.

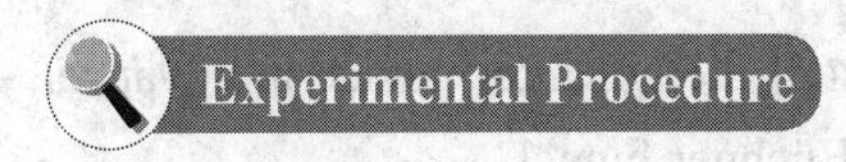

Experimental Procedure

Purification of Acetanilide——Recrystallization with 33% Ethanol as Solvent.

Put 2g of crude acetanilide into a 50ml round-bottom flask, then add 20ml of 33% ethanol and 2 grains of zeolite. Install reflux condenser tube (reflux apparatus as shown in Figure 2-3), then heat the mixture to boiling in a water bath. Control reflux speed by 1~2 drops per second. Shake the flask to accelerate the dissolution. Continue to add solvent from the upper end of the condenser tube if the acetanilide cannot be completely dissolved. When it is completely dissolved, add a few milliliters of extra solvent. Remove the heat source, when the mixture is cooled slightly, remove condenser and add an appropriate amount of activated carbon. Install reflux condenser again and reflux, heat and boil the mixture for 5 minutes. Stop heating, filter the mixture while it is hot, transfer the filtrate into a 50ml conical bottle. And cover the bottle mouth with a watch glass with convex face up. Cool it naturally to room temperature and then put it in ice water. Acetanilide will crystallize itself. Filter, collect crystals and wash it with a small amount of 33% ethanol. Drain the solvent, take out filter cake, dry it under infrared light, weigh, and calculate the yield. Observe the crystal form and color.

Preview Guide

Preview Requirements

1. Understand the principle and significance of recrystallization.
2. Learn the experimental method of recrystallization.

Notes

1. The purpose of adding activated carbon in hot saturated solution is to adsorb colored impurities. It cannot be directly added to boiling solution to avoid sudden boiling. The solution should be cooled slightly and then heated again to fully decolorize.
2. In addition to the use of thermal filter funnel, suction filter can be used for hot filtration, if suction filtration is used, 20%~30% more solvent should be added.
3. Natural cooling method is mostly used for crystallization.
4. We should properly control the drying temperature of the product, otherwise the solid will melt.

Experimental Explanation

1. When adding solid samples to the round-bottom flask, be careful not to stick the solid samples to the grinding mouth, otherwise after heating and reflux, the round-bottom flask and the condensing tube will stick together and cannot be separated.
2. Filter paper size should be cut properly, otherwise it will introduce impurities such as activated carbon mixing with filtrate.
3. The filter paper needs to be wetted in advance, otherwise it will cause impurities such as activated

carbon to be mixed into the filtrate.

4. Prepare the filter paper and hot solvent in advance, and install the hot filter device in advance, otherwise the impurities and crystals will precipitate together on the Büchner funnel.

5. In the process of washing the crystal, do not wash while filtering, otherwise the solvent will not have time to contact with the crystal and be removed.

6. In the process of washing the crystal, make sure that the filter paper is not loose, otherwise the crystal will leak into the filter bottle.

Questions

1. What steps do recrystallization operations include? What is the purpose of each step?

2. Why shouldn't the amount of solvent be too much or too little during recrystallization? How should we determine the correct amount of solvent?

3. When refluxing, can the upper opening of the condenser tube be plugged with a plug?

实验三 重结晶

实验目的

1. 掌握重结晶的原理和实验方法。
2. 掌握过滤、回流等基本操作。

实验原理

有机反应所得的产物，都混有杂质，如未反应的原料和副产物等。为了要得到比较纯净的物质，必须通过精制。一般有机物都是以固体或液体两种形态存在，它们的精制方法也各不相同。纯化固体最为常用的方法就是重结晶，对于容易升华的物质则往往采取升华来提纯。

视频

所谓重结晶，其一般过程就是设法将固体物质溶解于某种适当的溶剂中，溶液经过过滤（必要时须经脱色）除去杂质后再经浓缩，冷却或其他方法处理，使较纯的结晶析出。过滤后所得的滤液为母液，所得的结晶即为纯化物质，有时这种操作需要反复进行，方能获得纯品。

1. 溶剂的选择 在进行重结晶时，选择合适的溶剂是一个关键问题。所选溶剂不能与重结晶物质发生化学反应，重结晶物质与杂质在溶剂中的溶解度有比较大的差别，一般重结晶物质随温度的不同，溶解度有显著的不同，即在热时容易溶，冷时难溶或不溶。且溶剂容易挥发，容易与重结晶物质分离。此外也要适当考虑溶剂的毒性、易燃性和价格等。

重结晶常用的溶剂有水、乙醇、甲醇、乙醚、石油醚、冰醋酸和苯等，为了选择合适的溶剂，必须考虑被溶物质的成分和结构，除须查看化学手册外，有时需要做溶解度试验。

溶解度试验方法：取 0.1g 待重结晶固体，放入一小试管中，用滴管将某一溶剂逐滴加入，不断振摇试管，当加入溶剂量接近 1 毫升时，加热混合物使沸腾（注意溶剂的易燃性），如此物质易溶于沸腾的溶剂中，即表示该溶剂不适合。如果样品不溶于沸腾的溶剂，则可逐步添加溶剂，每次约 0.5ml，并继续加热使沸腾，如加入溶剂量达 3ml，而物质仍不溶解，表示此溶剂也不适用。如该物质能溶解于 3ml 以内的热溶剂中，则将试管进行冷却，观察有无结晶析出，必要时可用玻璃棒摩擦试管内壁，以帮助晶体析出。如果浸入冷却液中，并用玻璃棒摩擦试管内壁后，在数分钟内仍无晶体析出，表示该溶剂不适于单独作溶剂。

在许多干燥试管中，按上述方法逐一采用不同的溶剂进行试验，观察在哪一种溶剂中冷却析出结晶最多则是最适用的溶剂，然后可将它与样品按适当的比例进行重结晶。

在不能选择到一种适当的溶剂时，一般采用混合溶剂。这是一对可互溶的溶剂，进行结晶的化合物在其中一种溶剂中可溶，而在另一溶剂中难溶。常用的混合溶剂有水 – 乙醇、水 – 乙酸、水 – 丙酮、水 – 吡啶、石油醚 – 苯、石油醚 – 乙醚、石油醚 – 丙酮、乙醚 – 乙醇等。

2. 加热 重结晶往往需要制备样品的饱和溶液，如果用水作溶剂，可用普通敞口容器加热。而如果用有机溶剂，则需用回流加热法（图 2-3），以防溶剂挥发损失及燃烧爆炸等危险。

3. 过滤 重结晶中，常用的过滤有趁热过滤和抽气过滤。

（1）趁热过滤　为了防止主要溶质在过滤过程中析出，所以需要趁热过滤。一般采用热水漏斗，将它固定安装妥当后，预先将夹套中的水烧热，然后过滤，如图 2-4。如溶液量少，也可选一短而粗颈的玻璃漏斗放在烘箱中预热，过滤时趁热取出过滤。热过滤时为加快过滤速度，常使用折叠式滤纸，折叠方法如图 2-5 所示：取一圆形滤纸对折再对折，然后在两折痕间不断对折，最后打开得到均匀的 16 等分的扇形，即得。

（2）抽气过滤　简称抽滤，可加快过滤的速度，同时还可使晶体与母液尽量分开，一般常用布氏漏斗与吸滤瓶（晶体量少于 0.5g 时用玻璃钉过滤），装置见图 2-6，抽气动力常用真空水泵。

4. 析出结晶　主要方法有两种。其一是减少溶剂的体积（浓缩），方法可用自然蒸发或加热蒸发；其二是降低溶液的温度。如果所选用的溶剂，不仅对杂质的去除有一定的作用，同时对样品的溶解度也能因温度的变化而有显著的差别（例如样品溶于热溶剂中而难溶于冷溶剂中），这样便可将样品溶于最小量的沸腾的溶剂中，趁热过滤除去不溶性的杂质，放冷后便有较纯的结晶析出。

冷却的方法对结晶的大小与产品的纯度有一定影响。若迅速搅拌冷却得到的晶体细小，表面积大，杂质易吸附在晶体表面；而缓慢冷却得到的晶体颗粒较大，往往母液和杂质易包在晶体内部。要得到纯度较好的晶体，要视杂质情况采用合适的冷却方法。

5. 晶体的洗涤和干燥　晶体抽滤后，为了使结晶与母液尽量分开，在抽滤时，应充分地洗涤，一般多在布氏漏斗上进行，即抽干母液后，停止抽气，用少量的洗涤液倒在沉淀上，然后以玻璃棒搅动，充分地洗涤（注意：不可搅破滤纸）再抽去洗涤液，如此操作 2~3 次，洗涤次数可多几次，每次洗涤必须抽干后再洗第二次，否则不容易洗净。

把彻底洗净压干的沉淀连带滤纸一起从漏斗上取出，放在干净的玻璃表面皿上，用镊子或匙揭去湿滤纸（注意：不可将滤纸上的纤维刮下来）进行干燥，常用的方法是在水浴上、电烘箱中加热，或用红外线照射干燥，也可放在干燥器中干燥。

仪器和试剂

1. 仪器　50ml 锥形瓶、250ml 烧杯、三角漏斗、滤纸、布氏漏斗、吸滤瓶、玻棒、表面皿、水泵、干燥箱、铁架台、铁圈、50ml 圆底烧瓶、球形冷凝管。

2. 试剂　乙酰苯胺粗品、活性炭、33% 乙醇。

实验步骤

乙酰苯胺的精制——以 33% 乙醇为溶剂的重结晶

称取 2g 乙酰苯胺粗品，置于 50ml 圆底烧瓶中，加入 20ml 33% 的乙醇和 2 粒沸石。装上回流冷凝管（回流装置如图 2-3），在水浴上加热至沸，控制回流速度 1~2 滴 / 秒，振摇加速溶解。若乙酰苯胺不能全部溶解，则从冷凝管上端继续加溶剂，加热振摇促进溶解，待完全溶解后，再多加几毫升溶剂。移除热源，稍冷后取下冷凝管，加入少许活性炭，再装上冷凝管回流，加热煮沸 5 分钟。趁热抽滤，滤液尽快转移入 50ml 锥形瓶，瓶口盖一表面皿（凸面朝上），自然冷却至室温后，再用冰水冷却。抽滤，收集晶体，并用少量 33% 的乙醇洗涤，抽干，取出滤饼，在红外灯下干燥，称重，计算收率。观察晶体颜色与形状。

预习指导

1. 理解重结晶的原理及意义。
2. 学习重结晶实验方法。

注意事项

1. 在热饱和溶液中加活性炭的目的是为了吸附有色杂质，不能直接加到沸腾的溶液中，以免暴沸，应将溶液稍冷却后再加，然后再继续加热以充分脱色。
2. 趁热过滤除了可以用热滤漏斗，也可用抽滤，如果用抽滤，应多加20%~30%的溶剂。
3. 冷却析晶，大多用自然冷却法。
4. 要适当控制产品的干燥温度，否则固体将会熔化。

实验说明

1. 在圆底烧瓶中加入固体样品时，小心不要将固体样品黏附在磨口处，否则加热回流后，导致圆底烧瓶与冷凝管粘连在一起，无法分开。
2. 滤纸大小裁剪不合适，导致活性炭等杂质混入滤液中。
3. 滤纸需提前润湿，否则会导致活性炭等杂质混入滤液中。
4. 事先要准备好滤纸、热溶剂等，提前安装趁热抽滤装置，否则会导致杂质和晶体一起在布氏漏斗上析出。
5. 洗涤晶体过程中，不能边抽滤边洗涤，否则会导致溶剂还没来得及与晶体接触而被抽走。
6. 洗涤晶体过程中，不可使滤纸松动，否则会导致晶体漏入抽滤瓶。

思考题

1. 重结晶操作包括哪几个步骤？每一步骤的目的各是什么？
2. 为什么重结晶时溶剂的用量既不能过多也不能过少？应如何确定正确的溶剂用量？
3. 回流时，冷凝管上口能否用塞子塞住？

4 Simple Distillation

Experimental Purpose

1. Understand the principle of simple distillation (蒸馏).
2. Master the operation method of simple distillation.
3. Be familiar with simple distillation application.

Experimental Principle

The most commonly used method of purifying liquids is distillation, a process that consists of vaporizing the liquid by heating and condensing the vapor in a separate vessel to yield distillate (馏出物). When a pure liquid is distilled, vapor rises from the distilling flask and contacts with a thermometer. The vapor then passes through a condenser (冷凝器) which liquefies (液化) the vapor and passes it into the receiving flask. The temperature observed during the distillation of a pure substance will remain constant throughout the distillation as long as both vapor and liquid are present in the system. When two components which have a large boiling point difference are distilled, the temperature remains constant while the first component distills. If the temperature remains constant, a relatively purer substance is being distilled. After the first substance distills, the vapor temperature will rise again, and then the second component will distill, again at a constant temperature.

Simple distillation can sometimes be used to separate a mixture of liquids with more than 30℃ difference in boiling points. When boiling points have little difference, or a high purity is desired, it is necessary to do a fractional distillation (分馏).

In the distillation process, the boiling point range of pure liquid organic compound is very narrow (0.5~1℃). Therefore, distillation can be used to determine the boiling point. This method is called macro method (常量法). In the macro method, the amount of reagent should be more than 10ml. If there is not so many samples, you may use micro method (微量法).

Instruments and Reagents

1. Instruments The distillation unit mainly consists of three parts: distillation, condensation and liquid receiving.Distillation apparatus mainly includes: a round-bottom flask, a distillation head (蒸馏头), a thermometer, a thermometer adapter (温度计套管), a condenser, a distillation adapter (尾接管), a conical flask (锥形瓶) or round-bottom flask.

2. Reagents industrial ethanol , *N, N*-dimethylformamide.

 Experimental Procedure

1. Distillation of industrial ethanol 40ml of industrial ethanol and 1~2 pieces of boiling chips (沸石) are added into a 100ml round-bottomed flask. Assemble the apparatus for simple distillation according to Figure 2-7. Use an appropriate size receiving flask to collect the anticipated volume of distillate. The position of the thermometer bulb is particularly important: the top of the bulb should be at the same level with the bottom of the sidearm of the distillation head. Have the assembly checked by your instructor. Turn on the water tap and adjust the water flow through the condenser to a proper flow rate. Heat the round-bottom flask under the suggestion by your instructor. As soon as the liquid begins to boil and the condensing vapor has reached the thermometer bulb, regulate the heat supply so that distillation continues steadily at a rate of 1~2 滴 / 秒 . When the distillation rate is adjusted and the thermometer has reached a steady temperature, change the receiving flask, read and record the head temperature. Continue the distillation, periodically record the head temperature until only 2~3ml of industrial ethanol remains in the round-bottom flask, and then discontinue heating, stop the condensate water, and disassemble the apparatus. Fractions (portions collected in separate flasks) are collected over narrow ranges of temperature. Read and record the distillation range of industrial ethanol that you have observed. Record the volume of distilled industrial ethanol that you obtain.

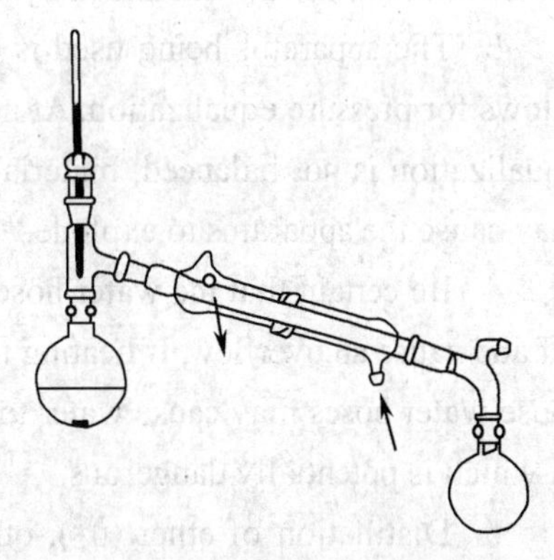

Figure 2-7 Simple Distillation Apparatus
图 2-7 简单蒸馏装置

2. Distillation of *N*, *N*-dimethylformamide (*N*, *N*- 二甲基甲酰胺，DMF) 20ml of DMF and 1~2 pieces of boiling chips are added into a 50ml round-bottomed flask. Use air-cooled condenser (空气冷凝管), the apparatus and operation are the same as distillation of industrial ethanol.

Preview Guide

 Preview Requirements

1. Understand the principle and application range of simple distillation.
2. Learn the assembly and disassembly sequence and operation method of simple distillation units.
3. Explain the relationship between boiling range and liquid purity.
4. Compare and master the similarities and differences of distillation apparatus and operations from the physical properties of ethanol and DMF.

 Notes

1. Always examine your glassware for cracks or any other flaws before assembling. Look particularly for "star crack" in round-bottomed flasks because these may cause the flask to break while

being heated.

2. Proper assembly of glassware is important to avoid possible breakage, spillage or release of distillate vapors into the room. Be certain that all connections in the apparatus are tight before beginning the distillation. Have it examined by your instructor after it is assembled.

3. The apparatus being used is open to the atmosphere at the receiving end of the condenser. This allows for pressure equalization. At no time in the laboratory a close system should be heated. If pressure equalization is not balanced, material expansion within the system will result in elevated pressures and may cause the apparatus to explode.

4. Be certain that the water hoses are securely fastened to your condensers so that they will not pop off and cause an overflow. If heating mantles (电热套) or oil baths are used for heating in this experiment, loose water hoses may cause water to spray onto electrical connections or into the heating sources, either of which is potentially dangerous.

5. Distillation of ether (醚), other low-boiling point liquid or toxic liquid, a rubber tube can be connected in the tail pipe branch, leading into sink or to outdoor (tail gas absorption device has to be installed for toxic gas) to avoid excessive inhalation (吸入) of organic vapors at all times.

6. Distillation must be stopped before the round-bottom flask becomes completely dry. Without the absorption of heat by vaporization, the flask temperature can rise very rapidly. Many liquids, particularly alkenes (烯烃) and ethers, may contain peroxides (过氧化物) which are easily explosive when concentrated in high residues.

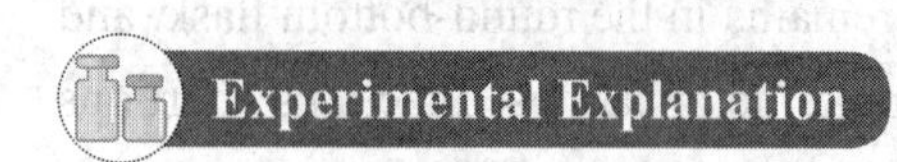

Experimental Explanation

1. Low-boiling organic liquids (for example ethyl ether) are highly flammable, so be sure that burners are not being used in the laboratory. Use heating mantle or water bath.

2. When distilling materials that have boiling points higher than 130°C, use an air-cooled condenser.

3. Choose a round-bottom flask that will leave 1/3~2/3 empty of the bulb.

4. Note the location of the “water in” and “water out” hoses on the condenser. The tube carrying the incoming water is always attached to the lower point, which ensures that the condenser is always filled with water. Do not create a fast flow or the increased pressure in the apparatus that may cause a piece of rubber tubing to pop off, spraying water everywhere. Use a fast flow of condensing water only when necessary, and use metal wire to fasten the hoses to the condenser and the water faucet to minimize the danger of popping off.

5. If forget to add boiling chips when the liquid is already being heated, you must cool the flask to below the boiling point before re-adding boiling chips. Because at this stage, the liquid might have already been overheated, it may erupt if boiling chips are added at high temperature.

6. When a distillation is interrupted, the pores on the boiling chips will be filled with liquid. Fresh boiling chips must be added before the distillation is continued.

Questions

1. What is the purpose of adding boiling chips?

2. For the condenser used in an apparatus for simple distillation, why should the lower opening be used for the water inlet?

3. Where should the bulb of the thermometer be placed? Explain how the location of the thermometer bulb influences the temperature reading.

4. Why can the distillate only occupy 1/3~2/3 of the flask?

5. Why is it dangerous to heat an organic compound in a distilling assembly that is closed tightly at every joint and has no vent or opening to the atmosphere?

实验四　常压蒸馏

实验目的

1. 理解常压蒸馏的原理。
2. 掌握常压蒸馏的操作方法。
3. 熟悉常压蒸馏的应用。

实验原理

视频

蒸馏是提纯液体物质的常用方法。蒸馏就是将液体物质加热到沸腾变为蒸汽，然后将蒸汽冷凝为液体的过程。蒸馏纯液体时，蒸汽从烧瓶中升起，触及温度计，再经过冷凝管，冷凝成液体，进入接受瓶。只要气液两相共存，温度将保持不变。蒸馏沸点相差较大的二组分液体，第一馏分蒸出时，温度保持不变。温度不变，馏出液纯度较高。第一馏分蒸完，温度再次上升，然后第二馏分蒸出，温度又保持不变。

沸点相差 30℃ 以上，可用普通蒸馏的方法分离。当二组分沸点相差不大，或需要高纯度，可用分馏法提纯。

纯液态有机化合物在蒸馏过程中沸点范围很小（0.5~1℃），所以，可以利用蒸馏来测定沸点，用蒸馏法测定沸点称为常量法，此法样品用量较大，需 10ml 以上，若样品不多时，可采用微量法。

仪器和试剂

1. 仪器　蒸馏装置主要包括三部分蒸馏、冷凝、接液。蒸馏装置主要包括圆底烧瓶、蒸馏头、温度计、温度计套管、冷凝管、接液管、锥形瓶或圆底烧瓶。

2. 试剂　工业乙醇、*N*，*N*– 二甲基甲酰胺。

实验步骤

1. 工业乙醇的蒸馏　在 100ml 圆底烧瓶中，加入 40ml 工业乙醇和 1~2 粒沸石，按图 2-7 安装实验装置，预测蒸出液体的体积，选用适当大小的接受瓶。温度计安放位置很重要，水银球的上限应和蒸馏头侧管口的下限在同一水平线上。指导教师检查装置后，打开冷凝水，调节到中等流速，按指导教师建议的方法加热圆底烧瓶，当液体开始沸腾、蒸汽到达包围温度计水银球时，调节加热速度，控制蒸馏速度 1~2 滴 / 秒。当蒸馏速度调好，温度计读数恒定时，更换接受瓶，观察并记录馏出温度。继续蒸馏，定期记录馏出温度，直到圆底烧瓶内仅剩 2~3ml 工业乙醇，停止加热，关闭冷凝水，拆除仪器。记录工业乙醇沸程，记下馏出液体积。

2. *N*, *N*– 二甲基甲酰胺（DMF）　在 50ml 圆底烧瓶中，加入 20ml DMF 和 1~2 粒沸石，用空

气冷凝管，装置操作同工业乙醇蒸馏。

预习指导

预习要求

1. 理解常压蒸馏的原理及其应用范围。
2. 学习常压蒸馏装置的装拆顺序和操作方法。
3. 说明沸程和液体纯度的关系。
4. 学会从乙醇和 DMF 的物理性质比较其蒸馏仪器和操作的异同点。

注意事项

1. 安装玻璃仪器前，要检查有无裂缝和其他缺陷，特别是检查圆底烧瓶有无星状裂纹，因为有裂纹的烧瓶加热时可能破损。

2. 正确安装玻璃仪器可以避免仪器的破损，液体溢出，蒸气泄漏。蒸馏前要确认接点紧密，装好仪器后，指导教师检查。

3. 接液管尾部与大气相通，内外压平衡。在实验室中，任何时候都不能加热密闭体系。如果内外压不平衡，系统内物质膨胀，压力上升，可能导致爆炸。

4. 确认导水管牢固的连接在冷凝管上，以防脱落造成冷凝水溢出。如使用电热套或油浴加热，水管松脱，水可能溅在电插头上或进入加热源，造成潜在危险。

5. 蒸馏乙醚等低沸点液体或有毒液体时，可在接液管支管处连接一橡胶管，引入水槽或通向室外（有毒气体需加尾气吸收装置），避免吸入过多有机蒸气。

6. 圆底烧瓶中液体完全蒸干前，应停止加热。若没有蒸发吸热，瓶温会迅速升高。许多液体特别是烯、醚可能含有过氧化物，浓缩后容易引起爆炸。

实验说明

1. 低沸点有机液体（如乙醚）易燃，实验室不能有明火，需采用电热套或水浴加热。

2. 蒸馏物质沸点超过 130℃，要使用空气冷凝管。

3. 圆底烧瓶大小的选择，蒸馏液体积占瓶容积的 1/3~2/3 为宜。

4. 注意冷凝管进出水位置，进水管位置低，保持冷凝管充满水。水流速度不可过快，过快水压会增高，可能导致橡胶管脱落使冷凝水溢出。只有在必要时，才用大流量冷凝水，这时需要用金属丝将橡胶管固定在冷凝管和水龙头上，降低松脱危险。

5. 如液体加热后，发现忘记加沸石，液体可能已经过热，烧瓶温度必须降到沸点以下，才能补加沸石，否则可能引起暴沸。

6. 如果蒸馏中断，液体不再沸腾，沸石上的小孔将充满液体，如需继续蒸馏，必须加入新的沸石。

思考题

1. 加沸石的目的是什么？
2. 为什么常压蒸馏用冷凝管较低的引水头为进水口？
3. 温度计水银球应处于什么位置？温度计水银球位置对温度读数有什么影响？
4. 蒸馏液为什么要占蒸馏瓶容积的 1/3~2/3？
5. 蒸馏装置接点严密、与大气不相通，加热蒸馏瓶中的有机物时为什么会有危险？

5 Fractional Distillation

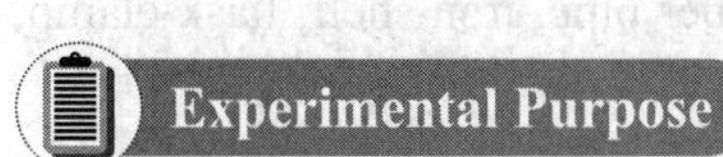

1. Understand the principle and significance of fractionation.
2. Master the operation of fractional distillation (分馏).
3. Know the application of fractional distillation.

Fractional distillation is applied to separating mutual soluble liquids with a difference in boiling points less than 25℃. It is more effective than simple distillation and is a commonly used technique of the separation and purification of liquid organic compounds with a smaller boiling point difference.

Fractional distillation is a distillation with a fractional column which is equal to several successive simple distillations. As the vapor from the distillation flask rises through the fractional column, it condenses on the column fillers and revaporises again. Every revaporization (再汽化) is just a separate distillation. Vapor and condensate are passing in opposite directions through the column. More volatile components rise through the column, while less volatile components flow down. The counter-flow is essential for effective separation in a fractional column. The vapor ascends through the column and condenses by heat exchange with the cooler fillers of the column. Heat exchange exists between vapor and condensate. This results in the more volatile components rising continuously in the fractional column. If repeated, that means the process goes through many vaporization-condensation cycles and has many distillations. So there are more volatile components near the top of fractional column and more high boiling-point components in the distillation flask. When the column efficiency is high enough, the low boiling-point component out of the top of the column and the high boiling-point component in the flask are purer. Thus, components with small differences in boiling point can be separated.

Fractional columns provide a large contact area to promote heat exchange and components separation. There are several common columns in Figure 2-8.

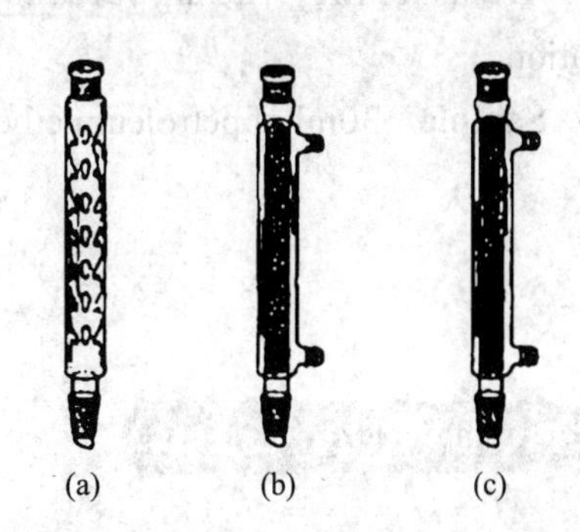

Figure 2-8 Common Fractional Columns

图 2-8 常用分馏柱

(a) spiny shaped (Vigreux) fractional column; (b) glass bead filled column; (c) stainless steel filled column

The fractional efficiency depends on the type of the fractional column, the height of the fillers and the adiabatic (绝热的) nature. The fractional column is often wrapped with asbestos (石棉) to achieve better effect.

To increase the contact area, the column is filled

with fillers as short glass tubes or glass beads. Space should be left among fillers to allow condensate and vapor to flow smoothly. To prevent the fillers from falling off, a few glass wires can be placed at the bottom of the column first.

Instruments and Reagents

1. Apparatus 100ml round-bottom flask, Vigreux fractional column, distillation head, straight condensate tube, thermometer, vacuum adapter, measuring cylinder, rubber pipe, iron shelf, flask clamp, condenser clamp, boiling chips, and electric heating jacket.

2. Reagent petroleum ether (bp 60~90℃).

Experimental Procedure

1. Select a suitable heat source, and assemble an apparatus of fractional distillation from left to right as shown in Figure 2-9. Add an appropriate amount of liquid to be fractionated and 2~3 pieces of boiling chips into the flask. Introduce tap water into the condenser. Start heating slowly so that the temperature of the distillates rises slowly and evenly.

2. When the liquid begins to boil and the vapor enters the column, the heat source temperature should be adjusted so that the condensed vapor ring rises slowly in the column. If the cold external wall makes the vapor rise too slowly, wrap the fractional column with asbestos.

3. When the vapor rises to the top of the column and the distillate appears, we should adjust the heat source temperature and control the speed of the distillate to 2~4 seconds per drop.

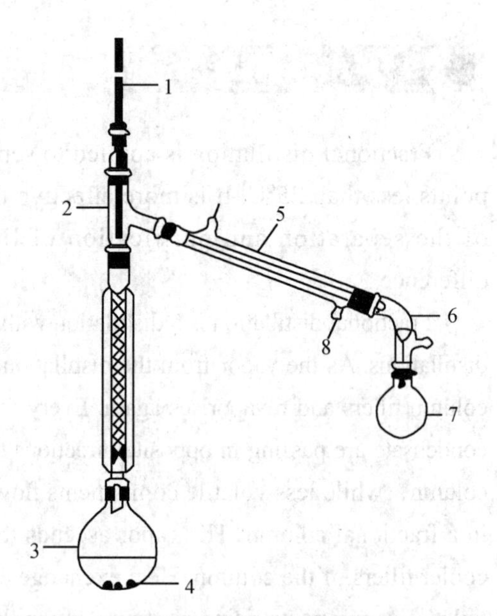

Figure 2-9 Apparatus for Fractional Distillation

图2-9 分馏装置图

1. thermometer; 2. distillation head; 3. round-bottom. flask; 4. boiling chips; 5. condenser; 6. vacuum adapter; 7. receiver; 8. water

4. Collect distillates at different temperatures respectively. Record the temperature and volume.

5. Disassembly the apparatus in turn after distillation.

6. Sample 30ml of petroleum ether.

Preview Guide

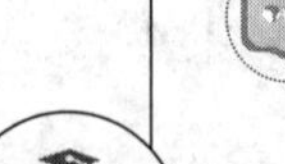

Preview Requirements

1. Understand the principle and application of fractional distillation.
2. Learn the assembly sequence and operation method of fractional distillation apparatus.
3. Learn the importance of temperature control in fractionation.

Notes

1. Control heating speed and let the vapor slowly rise to the top of the column.
2. If the distillation rate is too quick, the purity of products will decrease.
3. Wrap the column with asbestos when the outside temperature is low.

Experimental Explanation

1. At the end of fractionation, if no steam is around the mercury bulb of the thermometer, the temperature will drop.
2. Petroleum ether is easy to volatile, so it must be sealed after extraction and recovery.

Questions

1. Why does fractionation have better effect than simple distillation?
2. Why does heating fast reduce fractionation effect?

实验五 分馏

实验目的

1. 理解分馏的原理和意义。
2. 掌握分馏的操作方法。
3. 了解分馏的应用。

实验原理

视频

液体混合物中的各组分，若其沸点差小于25℃，则用普通蒸馏法难以精确分离，就应当用分馏的方法分离。分馏是比蒸馏更高效的分离沸点相近的液态化合物的常用方法。

分馏是使用分馏柱的蒸馏操作，相当于多次连续的简单蒸馏。当蒸馏瓶中的蒸气穿过分馏柱上升的时候，它遇到柱子中的填充物冷凝，然后再次蒸发。每次冷凝液的再蒸发相当于一次独立的蒸馏。蒸气和冷凝液在柱子中向相反方向运动，挥发性强的成分上升，挥发性弱的成分下行。反向流动对分馏柱中的有效分离非常重要。上升的蒸气遇柱内冷的填充物或者下行的冷凝液，进行热交换而部分冷凝。这样，上升蒸气中的易挥发组分含量越来越高，多次重复这一过程就能达到经过多次蒸发 - 冷凝循环的蒸馏效果。因此在分馏柱顶端，挥发性组分的含量很高，而高沸点的成分流回蒸馏瓶内。效率足够高时，从分馏柱顶端收集的蒸气接近纯的低沸点组分，同时，高沸点组分留在烧瓶内。这样，沸点相近的混合物得以分离。

分馏柱的作用是增加蒸气在柱内的冷凝面积，促使热交换，使之分离完全，常用的分馏柱见图2-8。

分馏效率取决于分馏柱的种类、填充料高度和绝热性质。为了达到较好的分馏效果，在操作时往往把分馏柱用石棉包起来。

为了增大与蒸气接触的面积，分馏柱中都装有填充料(如短玻璃管或玻璃珠等)；填充料间应留有空间，使冷凝液与蒸气有通畅对流的可能。为了防止填充料掉下来，分馏柱底部可放玻璃丝少许，再将填充料放置其上。

仪器与试剂

1. 仪器 100ml圆底烧瓶、韦氏分馏柱、蒸馏头、直形冷凝管、温度计、尾接管、量筒、乳胶管、铁架台、烧瓶夹、冷凝管夹、止爆剂、电热套。

2. 试剂 石油醚（bp 60~90℃）。

实验步骤

1. 选择好合适的热源，安装如图2-9，自左向右安装好装置后，烧瓶中加适量待分馏的液体，加入 2~3 粒沸石，通入冷凝水，即可开始加热。开始要缓慢加热，以便蒸馏液温度缓慢而均匀地上升。

2. 待液体开始沸腾，蒸气进入到分馏柱中时，要注意调节热源温度，使冷凝的蒸气环缓慢而均匀地沿分馏柱壁上升。如室温太低或液体沸点较高，分馏柱外壁散热太快，因而蒸气难于上升时，应将分馏柱用石棉包扎起来。

3. 当蒸气上升到分馏柱顶部，开始有馏出液流出时，更应密切注意调节热源温度，控制馏出液的速度为 2~3 秒/滴。

4. 收集不同温度下的馏分，记录温度和体积。

5. 蒸馏完毕后，依次拆除装置。

6. 实验样品　30ml 石油醚。

预习指导

预习要求

1. 理解分馏的原理及其应用范围。
2. 学习分馏装置的装拆顺序和操作方法。
3. 了解分馏过程中控制温度的重要性。

注意事项

1. 控制热源温度使蒸气缓慢上升到柱顶。
2. 馏出液流出速度太快，产品纯度将下降。
3. 若外界温度太低，可以包扎石棉给分馏柱保温。

实验说明

1. 分馏结束时，温度计水银球周围缺乏蒸气从而使温度下降。
2. 石油醚极易挥发，取用和回收完后及时加盖密封。

思考题

1. 为什么分馏比普通蒸馏分离效果好？
2. 为什么加热过快，分馏能力会下降？

PPT

6 Steam Distillation

Experimental Purpose

1. Understand the principle of steam distillation.
2. Master the operation method of steam distillation.
3. Know the application of steam distillation.

Experimental Principle

In steam distillation (水蒸气蒸馏), water steam is passed into the immiscible and volatile liquid in the distillation flask, and then the water and organic compound co-distill below 100℃.

The total vapor pressure is the sum of the partial pressures of water and the other insoluble component.

$$P=P_{H_2O}+P_A$$

In this formula, P is the total vapor pressure, P_{H_2O} is the saturated vapor pressure of water and P_A is the saturated vapor pressure of the insoluble component.

When the total vapor pressure achieves the atmospheric pressure, the mixture will start boiling at a lower temperature (boiling point) below either boiling point of the individual components. That means the organic component will co-distill out below its normal boiling point and below 100℃. (The amount of the components is n_{H_2O} and n_A respectively). The system temperature remains the same until one component is completely distilled out.

According to Dalton's law of partial pressure (道尔顿分压定律), the mole ratio of two components evaporated is

$$\frac{n_A}{n_{H_2O}}=\frac{P_A}{P_{H_2O}}$$

This formula is applicable to the calculation when component A is insoluble in water. In fact, any compound is partially dissolved in water. For the compound which is difficult to dissolve in water, the result obtained above is only an approximate value.

Steam distillation is a commonly method of extraction and separation. It is applicable to the following situations.

1. Mixture contains a large amount of solid or tar, which is hard to purify in any other way.

2. An organic compound is not or difficult to dissolve in water. Its vapor pressure is at least 5mmHg at 100℃.

3. A high boiling-point compound is easy to decompose, discolor or deteriorate at its boiling point and cannot react with water when co-boiling.

4. It is also suitable for the extraction of volatile oil from traditional Chinese medicine, such

as orange-peel oil from orange peel, peppermint (薄荷) oil from mint, Perilla oil from Perilla (紫苏属).

Instruments and Reagents

1. Instruments steam generator, 500ml round-bottom flask, 100ml conical flask, T-tube, straight condenser, vacuum adapter, air duct, electric furnace, asbestos net and rubber pipe.

2. Reagents Chinese medicine Cynanchumpaniculate, or peony bark, 2% $FeCl_3$ aqueous.

Experimental Procedure

1. Assemble an apparatus shown in Figure 2-10 from left to right successively. Be aware of the following points.

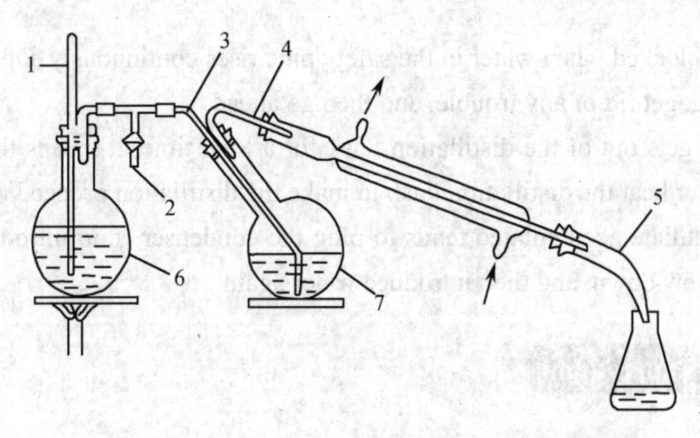

Figure 2-10 Typical Apparatus for Steam Distillation

图 2-10 水蒸气蒸馏装置图

1. safety pipe; 2. T-tube; 3. steam inlet pipe; 4. distillate outlet pipe; 5. adapter; 6. steam generator; 7. distillation flask

(1) Steam Generator Steam Generator is 1/2~3/4 full of water, with a safety pipe indicating internal pressure. The safety pipe should be inserted into water but not to the bottom.

(2) Distillation flask A round-bottom flask with suitable connectors with the sample to be distilled. The steam inlet pipe should be inserted close to the bottom of the bottle to make steam and the sample have a good contact.

2. Add the sample into the distillation flask. Make sure all the joints are tight. Introduce tap water into the condenser. Open the screw clamp on the T-tube and heat the steam generator. When the water boils, turn off the T-tube and let the steam enter the distillation flask. Adjust the heating source to keep the rate of distillate at 2~3 drops per second.

3. When the distillate is clear, the distillation finished. Open the screw clamp, stop heating and disassemble the apparatus in turn.

4. Sample Choose one sample from cyclohexane, peony bark and Cynanchumpaniculate fragments.

Preview Guide

Preview Requirements

1. Learn the principle and application of steam distillation.
2. Learn the installation sequence and operating method of steam distillation apparatus.
3. Explain the relationship between the saturated vapor pressure and the boiling point of liquid.
4. According to the structure, explain why paeonol can be extracted by steam distillation and tested by $FeCl_3$ aqueous.

Notes

1. The system is blocked when water in the safety pipe rises continuously. Open the screw clamp on the T-tube, stop heating, get rid of any trouble, and then go ahead.
2. If no distillate gets out of the distillation flask for a long time, it means the steam isn't enough. Increase the firepower or heat the distillation flask to make the distillation proceed more quickly.
3. If the solid distillate accumulated tends to plug the condenser, stop introduce condensate water, let the solid melt and flow down, and then introduce water again.

Experimental Explanation

Paeonol (or peony phenol, 4-methoxy-2-hydroxyacetophenone) is an effective component in Chinese medicine Cynanchumpaniculate. Paeonol has analgesic (镇痛的) effect and shows dark purple with $FeCl_3$ (See Experiment 26).

H_3CO—(benzene ring)—$\overset{O}{\overset{\|}{C}}CH_3$, OH

Questions

1. Why steam that is not hot enough must be let off?
2. If water in the safety pipe rises continuously, how to get rid of the trouble?
3. Why is the end of the steam tube as close to the bottom of the vessel as possible during the distillation process?
4. What is the condition of steam distillation? What conditions must be met for organic compounds used in steam distillation?

实验六 水蒸气蒸馏

实验目的

1. 理解水蒸气蒸馏的基本原理和应用范围。
2. 掌握水蒸气蒸馏的操作技术。
3. 了解水蒸气蒸馏的应用。

实验原理

水蒸气蒸馏，是将水蒸气通入不溶或难溶于水但在100℃时有一定挥发性的有机物质中，使需要蒸馏的物质在低于100℃的温度下随着水蒸气一起蒸馏出来。

当水和不（或难）溶于水的某化合物一起存在时，整个体系的蒸气压力为二者蒸气压之和，即

$$P=P_{H_2O}+P_A$$

式中 P 为体系蒸气压，P_{H_2O} 为水的饱和蒸气压，P_A 为不（或难）溶于水的化合物的饱和蒸气压。当 P 达到大气压时，体系开始沸腾，显然沸腾时的温度比水及该化合物的沸点都要低，也就是说该化合物和水在低于100℃时可被共同蒸出（二者物质的量分别为n_A和n_{H_2O}）。蒸馏时体系温度保持不变直至其中一组分被完全蒸出。

根据道尔顿分压定律，蒸出的两物质的量之比为

$$\frac{n_A}{n_{H_2O}}=\frac{P_A}{P_{H_2O}}$$

此式适用于当A物质在水中不溶解时的计算，实际上任何物质在水中都有部分溶解，对于难溶于水的物质，上式计算所得结果只是近似值。

水蒸气蒸馏是常用的提取分离方法，常适用于下列情况。

（1）混合物中含有大量固体或焦油状物质，通常其他方法不适用。

（2）混合物中存在不溶或难溶于水，而挥发性又较强的物质，该物质在100℃时蒸气压至少要有5mmHg。

（3）沸点很高，在接近或到达沸点时，容易分解，变色或变质，而在与水共沸时不发生化学反应。

（4）还适用于提取中药中的挥发油，如从陈皮中提取陈皮油，从薄荷里提取薄荷油，从紫苏里提取紫苏挥发油等。

仪器和试剂

1. 仪器 水蒸气发生器、500ml圆底烧瓶、100ml锥形瓶、T形管、直形冷凝管、引接管、

导气管、电炉、石棉网、乳胶管。

2. 试剂 中药徐长卿或牡丹皮、2% $FeCl_3$ 溶液。

实验步骤

1. 如图 2-10 安装，自左向右，依次安装好仪器装置。在安装中注意以下几项。

（1）水蒸气发生器 水蒸气发生器内装水量为容器容积的 1/2~3/4，附安全管以指示水位，并表示体系内压。安全管要插入水面以下但不能触底。

（2）蒸馏瓶 用圆底烧瓶配以合适的接头，内盛待蒸馏物质；导入蒸气的玻璃管应插入中央近瓶底，以便水蒸气与待蒸馏物质充分接触。

2. 将要蒸馏的物质装入蒸馏瓶内，确保各接口紧密接合。通入冷凝水，打开 T 形螺旋夹，加热水蒸气发生器。水沸后，关闭 T 形螺旋夹，使蒸气通入圆底烧瓶，开始蒸馏。调节加热速度，使馏出液的速度为 2~3 滴/秒。

3. 待馏液透明，表示已蒸完，此时应打开螺旋夹，停止加热，稍后关上冷凝水，逐步拆除装置。

4. 实验样品从环已烷、牡丹皮、徐长卿选其一。

预 习 指 导

预习要求

1. 学习水蒸气蒸馏的原理及其应用范围。
2. 学习水蒸气蒸馏装置的安装顺序和操作方法。
3. 说明液体的饱和蒸气压和沸点的关系。
4. 从丹皮酚的结构说明其能用水蒸气蒸馏提取及用 $FeCl_3$ 溶液检验的原因。

注意事项

1. 在蒸馏过程中，如安全管液面上升很快，表示有堵塞现象，则需打开排气管 T 形螺旋夹并停止加热，检查何处堵塞，待一切正常后再开始蒸馏。

2. 如果通入蒸汽一段时间后仍无液体蒸出，则说明蒸气量不足，烧瓶内液体温度太低，这时应调快加热速度，或用小火加热放置样品的烧瓶。

3. 如蒸出物为固体而有堵塞冷凝管的趋势，则可停止通入冷水片刻，待熔化后再慢慢通入冷水。

实验说明

丹皮酚（或牡丹酚，4- 甲氧基 -2- 羟基苯乙酮）为中药徐长卿和牡丹皮中的有效成分，具有镇痛作用，与 $FeCl_3$ 显示暗紫色。

思考题

1. 为什么要把不够热的水蒸气放出？
2. 如安全管中水位不断升高，应如何排除故障？
3. 进行水蒸气蒸馏时，蒸气导管的末端为什么要尽可能接近容器的底部？
4. 什么情况下可选择水蒸气蒸馏？水蒸气蒸馏的有机物必须满足什么条件？

7 Vacuum Distillation

Experimental Purpose

1. Master the operating technique of vacuum distillation.
2. Understand the principle of vacuum distillation.

Experimental Principle

The boiling point of liquid is the boiling temperature when its vapor pressure equals external pressure. The boiling temperature of liquid will increase along with the increase of external pressure, and decreases along with the decrease of external pressure. Many organic compounds, especially those with high boiling points, undergo partial or total decomposition in distillation at atmospheric pressure. In those cases, vacuum distillation (减压蒸馏) is the most effective method. When the pressure is reduced to 2.66 kPa (equal to 20mmHg), the boiling points of general high boiling point organic compounds are 100~120℃ lower than those under normal pressure.

Instruments and Reagents

1. Instruments 100ml and 50ml round-bottom flask, Claisen distillation head, capillary tube (毛细管), thermometer, straight condenser tube, vacuum-distilling adapter, rubber pipe, screw clamp, water bath pot, water pump.

2. Reagents anhydrous alcohol.

Experimental Procedure

1. The instrument is installed from left to right successively, as shown in Figure 2-11. Pay attention to the following points. The capillary must be inserted to 1~2mm from the bottom of the bottle, and the capillary must not be too thick, otherwise it will affect the distillation. Check the airtightness of distillation system to ensure that all parts of the apparatus are tightly connected and do not leak air.

2. Turn on the water pump and adjust the screw clamp above the safety bottle and the capillary tube to make the reading on the vacuum gauge reach the required value, and then heat with an appropriate hot bath.

3. Collect the low-boiling fraction into one of the receiving bottles. When the distillation temperature rises to the required temperature, rotate the grinding port of the receiving part so that the receiving liquid flows into another receiving bottle at a rate of no more than one drop per second. The distillation can be stopped until the temperature changes.

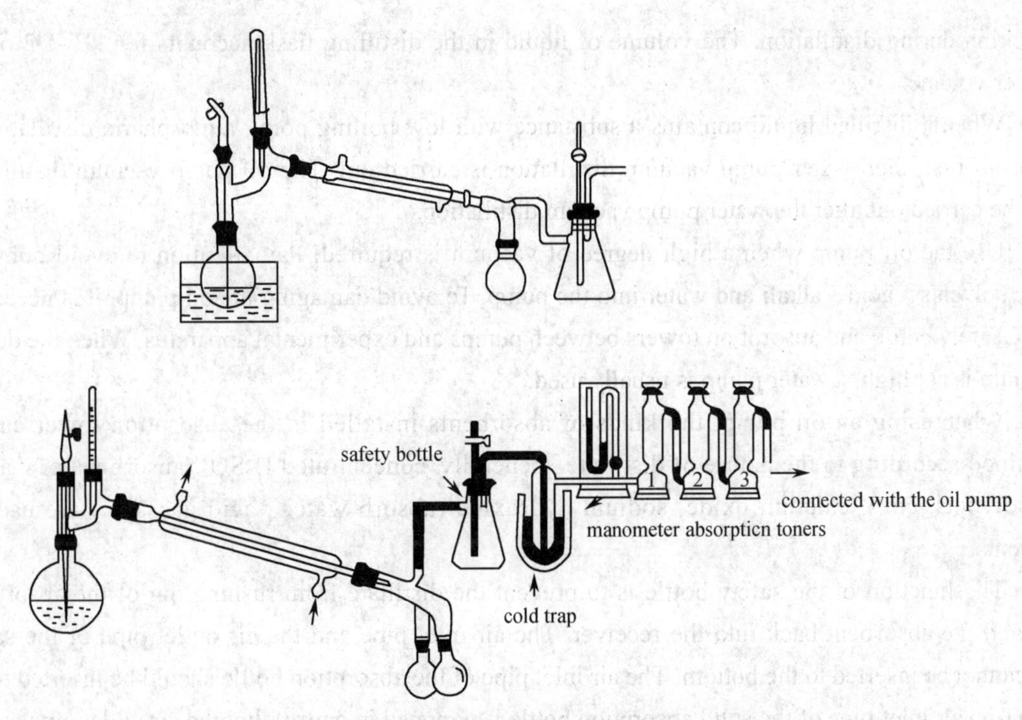

Figure 2-11 Typical Apparatus for Vacuum Distillation

图 2-11 减压蒸馏装置图

absorption towers: 1.anhydrous $CaCl_2$; 2.solid NaOH; 3.paraffin

4. At the end of distillation, first remove the heat source, then slightly loosen the piston of the capillary tube, followed by slightly loosening the piston of the safety bottle to fully open the piston of the capillary tube after being communicated with the atmosphere. Stop the pump and dismantle the device.

5. Experimental sample 30ml of absolute ethanol. Collect distillate at a pressure of 0.04MPa and read the boiling point at this pressure.

Reagent	Vacuum Degree/MPa	Boiling Point/°C	Recovered Ethanol/ml
30ml Alcohol			

Preview Guide

Preview Requirements

1. Learn the principle and application of vacuum distillation.
2. Compare the difference between vacuum distillation and simple distillation.
3. Know the function of each part of vacuum distillation apparatus.
4. Explain the relationship between boiling point and pressure.

Notes

1. All Apparatus must be hard without any cracks in order to avoid the danger of explosion caused

by cracking during distillation. The volume of liquid in the distilling flask accounts for 1/3~1/2 of the container volume.

2. When a distilled liquid contains a substance with low boiling point, atmospheric distillation is carried out first, then water pump vacuum distillation is carried out. The oil pump vacuum distillation should be carried out after the water pump vacuum distillation.

3. Use the oil pump when a high degree of vacuum is required. Pay attention to avoid pumping organic solvents、acid、alkali and water into the pump. To avoid damaging the oil pump, it is necessary to add a safety bottle and absorption towers between pumps and experimental apparatus. When the degree of vacuum is not high, a water pump is usually used.

4. When using an oil pump, the kinds of absorbents installed in the absorption tower can be determined according to the nature of distillate. Generally, concentrated H_2SO_4 (absorb water、alkali gas、solvents, etc.), calcium oxide, sodium hydroxide (absorb water、acid gas, etc.) are used as absorbents.

5. The function of the safety bottle is to prevent the distillate from rushing out of the absorption bottle and the absorbent back into the receiver. The air inlet pipe and the air outlet pipe of the safety bottle cannot be inserted to the bottom. The air inlet pipe of the absorption bottle should be inserted to the bottom (the air inlet pipe of the solid absorption bottle is wrapped in gauze), but the air outlet pipe cannot be inserted to the bottom.

Experimental Explanation

1. The capillary is mainly used to maintain a stable boiling, which also plays a certain role of stirring to prevent violent boiling liquid.

The method for testing the thickness of the capillary is to insert the capillary tube into a test tube containing a little ether and blow air into the tube. If small bubbles come out of the capillary, indicating that it is appropriate. If there are many bubbles, it means that the capillary is thicker. At this time, a rubber pipe can be connected to the top of the capillary tube, and it is clamped by a screw clamp, which is properly opened during distillation to regulate the air flow and make the boiling stable.

2. When a distilled liquid contains a substance with low boiling point, atmospheric distillation is carried out first, then water pump vacuum distillation is carried out. The oil pump vacuum distillation should be carried out after the water pump vacuum distillation.

3. In practical work, vacuum distillation is often used to recover solvents, such as ethanol. At this time high vacuum is often not required, so water vacuum pumps can be used. Currently, the commonly used apparatus is a circulating water vacuum pump. If there is no special need, the absorption tower can often be saved.

4. Vacuum distillation requires first pumping air and then heating, because the system is full of air. Part of the solution is vaporized after heating, and air extraction is carried out at this time. A large amount of gas has no time to condense and is directly absorbed into the vacuum pump to damage the pump and change the vacuum degree.

5. During the experiment, the temperature should be set reasonably. If the distillate speed is too fast, organic vapor will not be condensed in time and be absorbed into the pump body.

Questions

1. Is it necessary to add boiling chips in vacuum distillation?
2. After distillation, why do we need to open screw clamps of the capillary and the safety bottle before turning off the pump?

实验七 减压蒸馏

实验目的

1. 掌握减压蒸馏的操作技术。
2. 理解减压蒸馏的原理。

实验原理

视频

液体的沸点是指它的蒸气压等于外界压力时的温度，此时液体会沸腾。液体沸腾的温度会随着外压的增加而升高，也会随着外压的减小而降低。很多有机化合物，特别是高沸点的有机化合物，在常压下蒸馏往往发生部分或全部分解。在这种情况下，采用减压蒸馏方法最为有效，一般的高沸点有机化合物，当压强降低至2.66kPa (相当于20mmHg) 时，其沸点要比常压下的沸点低100~120℃。

仪器和试剂

1. 仪器 100ml和50ml圆底烧瓶、克氏蒸馏头、毛细管、温度计、直形冷凝管、引接管、乳胶管、螺旋夹、水浴锅、水泵。

2. 试剂 无水乙醇。

实验步骤

1. 安装如图2-11，自左向右，依次安装好仪器。在安装时注意毛细管必须插至距离瓶底1~2毫米处，毛细管不能过粗，否则将影响蒸馏。进行气密性检查，确保装置各部分连接紧密不漏气。

2. 开动水泵，调节安全瓶和毛细管上方螺旋夹，使真空表上的读数到所需要的数值，即可用适当的加热。

3. 将前馏分收集至其中一接收瓶，待蒸馏温度升至所需温度时，旋转接收部分的磨口，使接收液流入另一接收瓶，蒸馏速度为每秒钟1滴。直至温度发生变化，即可停止蒸馏。

4. 蒸馏结束时，先移去热源，然后稍旋开毛细管的活塞，再稍旋开安全瓶的活塞，使与大气相通后完全打开毛细管的活塞。停泵，拆除装置。

5. 实验样品，如30ml无水乙醇。收集压力0.04MPa下的馏出液，并读取此压力下的沸点。

试剂	真空度 /MPa	沸点 /℃	回收乙醇 /ml
30ml 乙醇			

预习指导

预习要求

1. 学习减压蒸馏的原理及其应用范围。
2. 比较减压蒸馏与常压蒸馏装置的不同点。
3. 了解减压蒸馏装置各部分作用。
4. 说明沸点和压力之间的关系。

注意事项

1. 仪器都必须是硬质的，而且没有任何裂缝，以免在蒸馏过程中发生破裂，引起爆炸的危险。蒸馏瓶内液体体积占容器容积的1/3~1/2。

2. 进行减压蒸馏时不能用直火加热，必须用热浴加热（加热均匀，防止爆沸）。

3. 在真空度要求很高时使用油泵，此时注意切忌将有机溶剂、酸、碱和水抽入泵体，以免损坏油泵，装置中的安全瓶和吸收塔就是起这一作用的。在真空度要求不高时，一般用水泵。

4. 使用油泵时，吸收塔内所装吸收剂的种类，可根据蒸馏液的性质而定，一般用浓 H_2SO_4（吸收水分、碱气、溶剂等）、氧化钙、氢氧化钠等（吸收水分、酸气等）作为吸收剂。

5. 安全瓶的作用是防止蒸馏液冲出吸收瓶以及吸收剂倒吸入接收器，安全瓶的进气管和出气管都不能插到底，吸收瓶的进气管要插到底（固体吸收瓶中进气管用纱布包好），而出气管不能插到底。

实验说明

1. 毛细管主要用以维持平稳的沸腾，同时又起一定的搅拌作用，这样可以防止液体暴沸。检验毛细管粗细的方法是将毛细管插在盛有少许乙醚的试管中，向管内吹入空气，有很小的气泡由毛细管冒出，表示合适。如气泡很多，说明毛细管较粗。此时可在毛细管最上端接一橡皮管，用螺旋夹夹紧，在蒸馏时适当打开少许，以调节空气流，使沸腾平稳。

2. 被蒸馏液体中若含有低沸点物质时，通常先进行普通蒸馏，再进行水泵减压蒸馏，而油泵减压蒸馏应在水泵减压蒸馏后进行。

3. 在实际工作中常可用减压蒸馏的方法来回收溶剂，如乙醇等，此时往往真空度要求不高，一般用水泵抽真空，现在常用的产品为循环水式真空泵，如果没有特殊需求，常可将吸收塔省去不用。

4. 减压蒸馏时需先抽气再加热，因为系统内充满空气，加热后部分溶液汽化，再抽气时，大量气体来不及冷凝和吸收，会直接进入真空泵，损坏泵改变真空度。

5. 实验过程中温度设置需合理，若馏出液速度过快，会导致有机物蒸汽来不及冷凝吸收进入泵体。

思考题

1. 减压蒸馏时是否需加止爆剂？
2. 减压蒸馏结束时为什么要先打开毛细管和安全瓶的活塞，然后才能关泵？

8 Extraction Using Separatory Funnels

Experimental Purpose

1. Understand the principle of liquid-liquid extraction.
2. Master the use method of separatory funnel.
3. Know the application of liquid-liquid extraction.

Experimental Principle

Extraction and washing are separate operations for different purposes based on the principle of compounds with different solubility in different solvents. If the compound extracted is needed, the operation is called extraction. If not, the operation is called washing. Here we simply call this operation extraction. Extraction is classified as liquid-liquid extraction and liquid-solid extraction, but usually refers to liquid-liquid extraction. In this experiment, liquid-liquid extraction is mainly introduced. For a solid mixture, liquid-solid extraction can be performed with a Soxhlet extractor (索氏提取器) (See Experiment 24).

The distribution law (分配律) is the principle for liquid-liquid extraction. The solubility of an organic compound is different in different solvents. For a solute in the system with immiscible (不混溶的) solvents A and B at a certain temperature (the solute does not decompose, ionize or associate, and has a certain solubility in both solvents), the concentration ratio in two solvents is a constant nothing to do with volumes of the two solvents and the amount of solute.

$$C_A/C_B=K$$

K is partition coefficient.

Organics are generally more soluble in organic solvents than in water, so those organics dissolved in water can be extracted by organic solvents. According to the distribution law, several smaller extractions are better than one big one. However, extraction for more than five times would improve little effect. Generally, three times achieve good effect.

Instruments and Reagents

1. **Instruments** 125ml separatory funnel (分液漏斗), 50ml beaker and 50ml conical flask.
2. **Reagents** paeonol aqueous solution, ethyl acetate and 2% $FeCl_3$ aqueous.

Experimental Procedure

1. Check the separatory funnel with some water. Make sure the stopper and the stopcock (旋塞) work well. If not, smear a little Vaseline (凡士林) on them.

2. Suspend a separatory funnel in an iron ring. Add the solution to be extracted and extractant. Replace the stopper. Pick up the funnel in both hands and invert it with the one hand holding the stopcock and the index finger of the other hand holding the stopper. Shake the funnel gently for a few seconds, vent the funnel slowly by opening the stopcock to release any pressure build-up. Close the stopcock and shake the funnel more vigorously, with occasional venting, so that the two immiscible solvents can contact each other efficiently. The operation is shown in Figure 2-12.

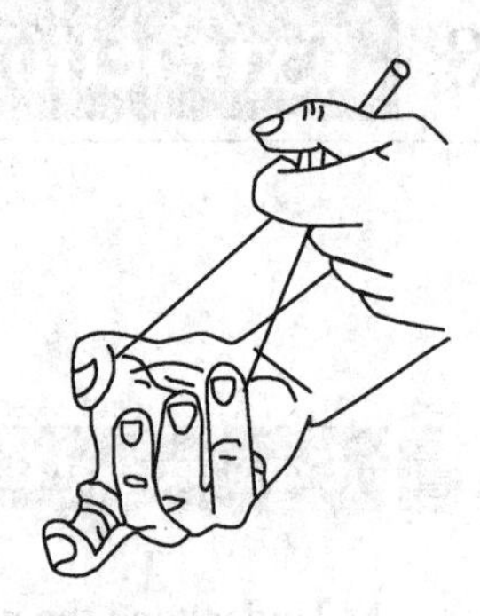

Figure 2-12 Venting a Separatory Funnel
图 2-12 分液漏斗放气

3. Place the funnel in a ring, and rotate the stopper, to make the solution communicate with atmosphere, and allow the funnel to stand until there is a sharp demarcation line (分界线) between the two layers. Drain the bottom layer into a conical flask by opening the stopcock. Adjust the stopcock to slow the drainage rate as the interface approaches the neck of the funnel. When the interface just reaches the neck, quickly close the stopcock to separate the layers cleanly. Pour the upper layer out from the top outlet.

4. If you get an emulsion (乳状液) after shaking, you do not have two distinct layers. Sometimes you can break up the suspended droplets by adding a little salt, or some acid or base. Be careful with the acid or base, and pay attention to whether it reacts with the extract.

5. Extraction of paeonol Extracting 20ml of paeonol aqueous with ethyl acetate for three times, 10ml each time. Add two drops of paeonol aqueous, the aqueous after first extraction and second extraction to the dropping plate, and add a drop of 2% $FeCl_3$ solution to the above three samples to compare the color.

Preview Guide

Preview Requirements

1. Understand the distribution law and the principle of liquid-liquid extraction.
2. Learn the request of extractant selection.
3. Be familiar with the use method of the separatory funnel.

Notes

1. Check the separatory funnel and make sure the stopper and the stopcock work well before experiment.

2. Don't let Vaseline adhere the inside wall of the funnel, otherwise it will contaminate organics by dissolving during extraction.

3. Drain the bottom layer into a conical flask by opening the stopcock. Pour the upper layer out

from the top outlet. If the upper layer is also drained through the stopcock, it will be contaminated by the residue of the bottom layer.

4. Retain both the upper and bottom layers until the experiment is finished. If something goes wrong, you can do it again.

Experimental Explanation

1. Shaking the mixture of ethyl acetate and paeonol aqueous in the funnel severely may cause emulsion. If emulsion occurs, add salt to demulsify it.

2. Use 2% $FeCl_3$ aqueous as blank control when testing paeonol on the spot plate.

Questions

1. How to select extractant?

2. Why is it necessary to vent gas in extraction operations?

实验八 分液漏斗萃取

实验目的

1. 理解液–液萃取的基本原理。
2. 掌握分液漏斗的使用方法。
3. 了解液–液萃取的应用。

实验原理

视频

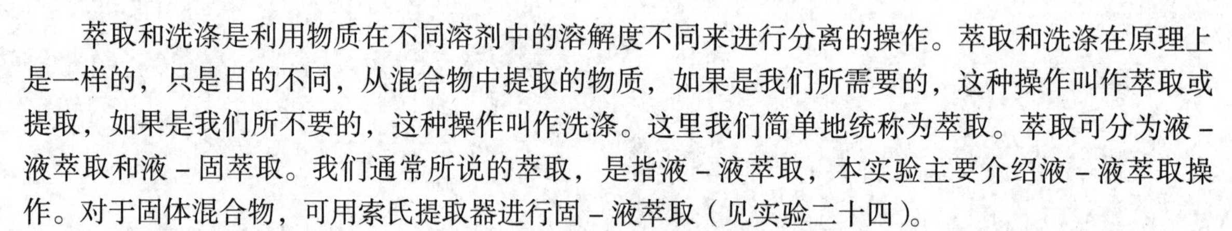

萃取和洗涤是利用物质在不同溶剂中的溶解度不同来进行分离的操作。萃取和洗涤在原理上是一样的，只是目的不同，从混合物中提取的物质，如果是我们所需要的，这种操作叫作萃取或提取，如果是我们所不要的，这种操作叫作洗涤。这里我们简单地统称为萃取。萃取可分为液－液萃取和液－固萃取。我们通常所说的萃取，是指液－液萃取，本实验主要介绍液－液萃取操作。对于固体混合物，可用索氏提取器进行固－液萃取（见实验二十四）。

分配定律是液－液萃取方法的主要理论依据。物质在不同溶剂中的溶解能力不同，如向互不相溶的两溶剂A和B共存的体系中，加入某物质（该物质在两种溶剂中均不发生分解、电离、缔合等，且在两溶剂中有一定的溶解度)。实验证明：在一定温度下，该物质在两溶剂中的浓度比是一个定值，而与两溶剂的体积及溶质的多少无关。

$$C_A/C_B=K$$

K 称为分配系数。

一般来说，有机化合物在有机溶剂中比在水中的溶解度大，常可用有机溶剂提取溶解于水中的有机物。根据分配原理，在萃取过程中，将一定量的溶剂分多次萃取，其效果要比一次萃取好，这就是我们常说的“少量多次原则”。不过若萃取次数大于5次，萃取效果提高甚微，一般3次就能达到良好的萃取效果。

仪器和试剂

1. 仪器 125ml 分液漏斗、50ml 烧杯、50ml 锥形瓶。

2. 试剂 丹皮酚水溶液、乙酸乙酯、2% $FeCl_3$ 溶液。

实验步骤

1. 向分液漏斗中加入适量水，然后检查分液漏斗的活塞和旋塞是否严密，旋塞转动是否灵活，如不灵活需涂凡士林，以防分液漏斗在使用过程中发生泄漏而造成损失。

2. 将分液漏斗悬置在铁圈上，倒入待萃取液与萃取剂。盖好活塞，开始振荡，先把分液漏斗倾斜倒置，一只手握住漏斗上口颈部，并用示指压紧活塞，另一只手握住旋塞。轻轻振荡漏斗几

秒钟后，打开旋塞放气。关上旋塞稍作剧烈振荡，再放气，让互不相容的两种溶剂充分接触，如图 2-12。

3. 把漏斗放置在铁圈上，旋转活塞与大气相通，静置漏斗至两液层完全分离。打开旋塞，用锥形瓶接收下层溶液。当液层界面快要到达分液漏斗颈部时，关小旋塞，控制下层流出液速度。液层界面到达颈部时，关闭旋塞，让两液层完全分离。上层液体从上口倒出。

4. 如果振荡后得到的是乳浊液，两液层不能完全分离。可以加入少量盐、酸或碱破坏乳浊液。但加酸、碱的时候要小心，注意是否与萃取物发生反应。

5. 萃取丹皮酚：丹皮酚水溶液 20ml 以乙酸乙酯萃取 3 次，每次 10ml。并分别取未经萃取的丹皮酚水溶液、第一次萃取和第二次萃取后的水层各两滴于点滴板上，加入 2% 的 $FeCl_3$ 溶液 1 滴，比较颜色深浅。

预 习 指 导

预习要求

1. 理解分配定律和液-液萃取的原理。
2. 学习萃取剂选择的要求。
3. 熟悉分液漏斗的使用方法及要点。

注意事项

1. 萃取前，一定要对分液漏斗检漏及检查活塞的灵活性。

2. 涂凡士林不能沾到分液漏斗内部，以免在萃取过程中，有机溶剂因溶解凡士林而被污染。

3. 下层液体从下口放出，上层液体从上口倒出。如果上层液体也经旋塞放出，则漏斗旋塞下面颈部所附着的残液就会将上层液体污染。

4. 在萃取或洗涤时，上下两层液体都应该保留到实验完毕。否则，如果中间的操作发生错误，便无法补救和检查。

实验说明

1. 萃取时乙酸乙酯和丹皮酚溶液不宜用力震荡，易产生乳化层。若发生乳化，可加入食盐去乳化。

2. 点滴板上加入 2% 的 $FeCl_3$ 溶液检验有无丹皮酚时，可用 2% 的 $FeCl_3$ 作空白对照。

思考题

1. 如何选择萃取剂？
2. 萃取操作中为何要放气？

Preparative Experiments

有机化合物合成实验

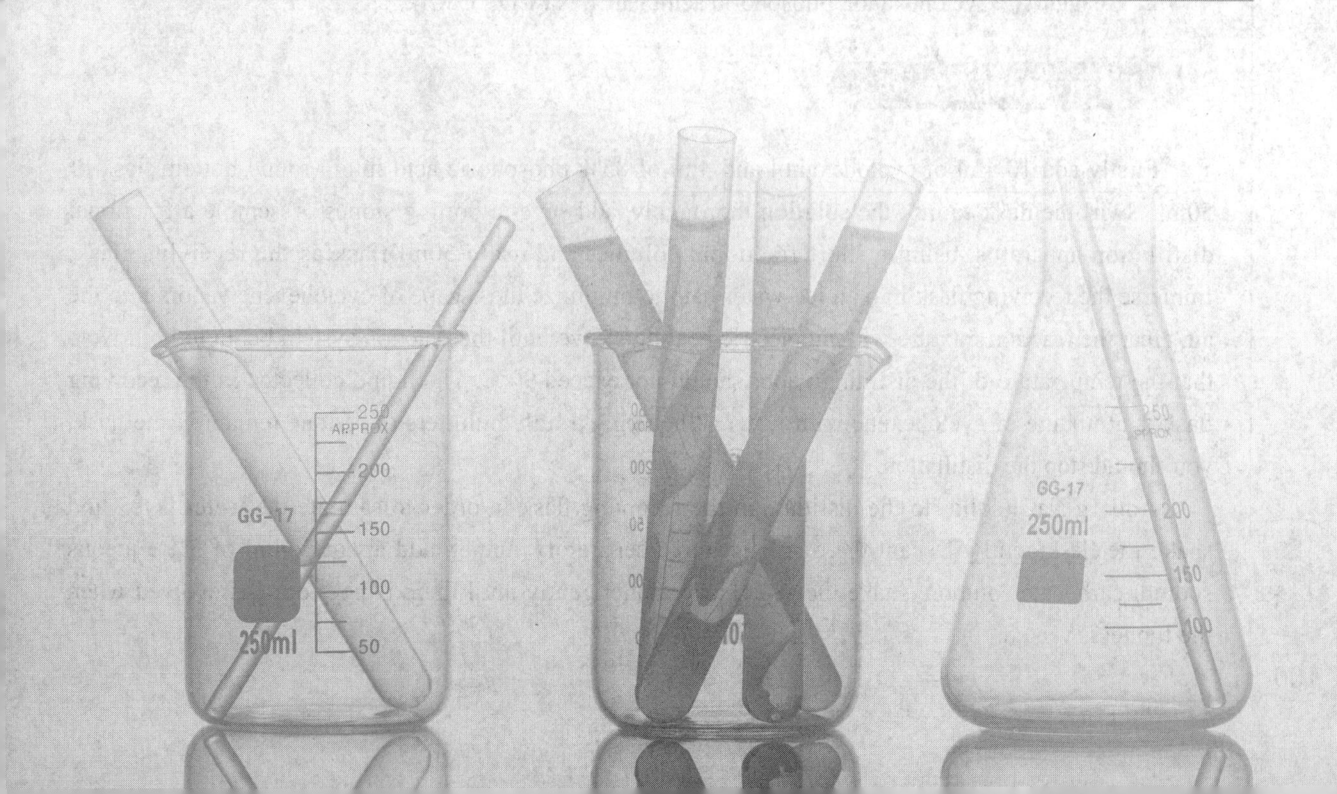

9 Synthesis of Cyclohexene

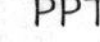

Experimental Purpose

1. Understand the principle and method of preparing cyclohexene (环己烯) by acid-catalyzed dehydration (消除) of cyclohexanol (环己醇).
2. Master the techniques of fractional distillation and simple distillation.

Experimental Principle

Cyclohexene is usually prepared by cyclohexanol dehydration in an acid catalyzed reaction using strong concentrated mineral acid such as phosphoric acid.

$$\text{cyclohexanol (OH)} \xrightarrow[\Delta]{H^+} \text{cyclohexene} + H_2O$$

Instruments and Reagents

1. **Instruments** Fractional distillation device and distillation instruments.
2. **Reagents** Cyclohexanol, phosphoric acid, NaCl, Na_2CO_3, $CaCl_2$.

Experimental Procedure

Firstly add 10.4ml of cyclohexanol and 4ml of 85% phosphoric acid into a round-bottom flask of 50ml. Swirl the flask to mix the solution thoroughly, add several boiling stones. Assemble a fractional distillation apparatus, using a short fractional column, and use a 50ml flask as the receiving flask. Immerse the receiving flask into an ice-water bath to minimize the escape of cyclohexene vapors into the air. Heat the reaction mixture with an electric heating sleeve until the reaction system begin to boil. Note that the temperature of the distilling vapor should not exceed 90℃. The liquid collected in the receiving flask is a mixture of cyclohexene with water. When only a little mililiters of residue remain in the flask, you should stop the distillation.

Add 1g NaCl(solid) to the distillate in the receiving flask in order to saturate the water layer, and shake the flask gently. Decant the distillate into a separatory funnel, add about 3~4ml of 5% aqueous sodium carbonate solution, shake the separatory funnel gently until there is no more gas evolved when the funnel is vented.

Firstly draw off the under aqueous layer, then pour the upper layer of crude cyclohexene through the upper neck of the separatory funnel into a small flask. Add about 1~2g of anhydrous calcium chloride to the flask, swirl the mixture occasionally until the total solution becomes clear. Decant this dry cyclohexene solution into a dry round-bottom flask of 50ml. Add several boiling stones and assemble the apparatus as a distillation device. Use a 50ml beaker flask submerged up to its neck in an ice-water bath as a receiver. Collect the fraction that boils between 80~85℃, then you can calculate the percentage yield.

Pure cyclohexene is a colorless liquid, b. p. 83℃, d_D^{20} 0.8102.

Preview Guide

Preview Requirements

1. Make good preparation by reviewing the theory and seeking information about the chemicals involved in preparation of preparing cyclohexene by acid-catalyzed dehydration of cyclohexanol, and pay attention to that cyclohexene and cyclohexanol will co-distill with the water.

2. Preview the techniques and list the important instruments of fractional distillation and simple distillation.

3. To explore the equipment drawing and list the main experimental procedure.

Notes

1. The equilibrium will be shifted in favor of the product, cyclohexene, by distilling it from the reaction mixture while it is being formed.

2. The cyclohexene (bp 83℃) will co-distill (共沸) with the water that is also formed (bp 70.8℃), and the starting material, cyclohexanol, also co-distills with the water (bp 97.8℃), so the distillation must be done carefully, not allowing the temperature to rise above 90℃.

3. The salt minimizes the solubility of the organic product in the aqueous layer.

4. As a small amount of phosphoric acid co-distills with the products, it must be removed by washing with aqueous sodium carbonate solution.

5. The solution should become clear, and if it is still cloudy, add additional 0.5g of drying agent, swirl and last for an additional 10min.

6. Make sure that all the glassware you will use for distilling is clean and dry. Dry all the glassware in an oven (110℃) for 10min.

Questions

1. Why must the temperature on the top of the fractional column during the reaction period be controlled?

2. What is the purpose of adding salt before the layers are neutralized and separated?

3. Why must any acid be neutralized with sodium carbonate before the final distillation of cyclohexene?

实验九 环己烯的制备

实验目的

1. 理解环己醇在酸催化下脱水制备环己烯的原理和方法。
2. 掌握分馏和简单蒸馏技术。

实验原理

环己烯可由环己醇在强的浓含氧无机酸催化下脱水制备，本实验以浓磷酸为脱水剂制备环己烯。

$$\text{环己醇(OH)} \xrightarrow[\triangle]{H^+} \text{环己烯} + H_2O$$

仪器与试剂

1. 仪器 分馏、蒸馏装置所需仪器。

2. 试剂 环己醇、浓磷酸、NaCl、Na_2CO_3、无水 $CaCl_2$。

实验步骤

在 50ml 圆底烧瓶中，加入 10.4ml 环己醇，4ml 85% 的浓磷酸和几粒沸石，充分振荡使之混合。安装分馏装置，使用短分馏柱，用 50ml 锥形瓶作接收器，并置于冰水浴中，以免环己烯挥发进入空气而损失。用电热套加热反应混合物至沸腾，控制分馏柱顶部蒸汽的温度不超过 90℃，慢慢蒸出生成的环己烯和水。当烧瓶中只剩下少量的残留物时，即可停止加热，全部加热反应时间约需 1h。

将馏出液用 1g 固体氯化钠饱和，并轻轻地振摇烧瓶。将馏出液倒入一个分液漏斗中，加入 3~4ml 5% 的碳酸钠溶液。轻轻振摇分液漏斗直至漏斗放气时无气体逸出。放出下层的水溶液层，上层的环己烯粗产物层从分液漏斗的上口倒入一个小锥形瓶中，加入 1~2g 无水氯化钙，振荡反应瓶，直至溶液变为澄清。将干燥后的环己烯滤入一个干燥的 50ml 圆底烧瓶中，加入几粒沸石，用水浴加热蒸馏，用一已称重的 50ml 锥形瓶作接收器，并将锥形瓶置于冰水浴中。

收集 80~85℃ 的馏分，称重产品，计算产率。纯的环己烯为无色液体，bp 83℃，d_D^{20} 0.8102。

预习指导

1. 查找环己烯制备实验所涉及的实验原理和理化参数，主要包括环己醇、环己烯，尤其注意它们与水形成共沸物的性质。

2. 预习实验技术，简要列出蒸馏和分馏装置所需仪器。

3. 试着画出反应装置图，列出主要实验步骤。

注意事项

1. 将生成的水和环己烯不断从反应体系中蒸馏出来，有利于反应平衡向产物方向移动。

2. 环己烯的沸点为83℃，环己烯与水共沸，沸点为70.8℃，原料环己醇也与水形成共沸物，沸点为97.8℃，因此，蒸馏温度需严格控制，使其不超过90℃。

3. 加氯化钠固体的目的是减少产物在水中的溶解度，达到更好地分离的目的。

4. 少量的磷酸也将与产物共沸，所以需用碳酸钠水溶液洗涤除去磷酸。

5. 干燥后的溶液应澄清，若仍浑浊，可补加0.5g干燥剂，振荡后再干燥10min。

6. 在精制产品时，蒸馏所用的仪器需在110℃的烘箱内干燥大约10min。

思考题

1. 在反应期间，为什么必须控制分馏柱顶部的温度？

2. 在液层被中和及分离前，加固体盐的目的是什么？

3. 产物环己烯被蒸馏前，为什么必须用碳酸钠将酸液调至中性？

10 Preparation of n-Bromobutane

Experimental Purpose

1. Understand the principle and method of preparing 1-bromobutane with *n*-butanol, concentrated sulfuric acid and sodium bromide.

2. Master reflux operations with hazardous gas absorption device.

Experimental Principle

1-Bromobutane (1-溴丁烷) is prepared by refluxing *n*-butanol with sodium bromide and concentrated sulfuric acid.

$$NaBr+H_2SO_4 \longrightarrow HBr+NaHSO_4$$

$$n\text{-}C_4H_9OH+HBr \xrightleftharpoons{H_2SO_4} n\text{-}C_4H_9Br+H_2O$$

Side reactions:

$$n\text{-}C_4H_9OH \xrightarrow[\triangle]{H_2SO_4} CH_3CH_2CH{=}CH_2+H_2O$$

$$2n\text{-}C_4H_9OH \xrightarrow[\triangle]{H_2SO_4} (n\text{-}C_4H_9)_2O+H_2O$$

$$2HBr+H_2SO_4 \xrightarrow[\triangle]{} Br_2+SO_2+2H_2O$$

Instruments and Reagents

1. Instruments 100ml round-bottom flask, spherical condenser, gas trap (气体分离器), distillation apparatus, separatory funnel, conical bottle and electric furnace.

2. Reagents *n*-butanol, concentrated sulfuric acid, sodium bromide, sodium hydroxide aqueous (5%), saturated sodium bicarbonate aqueous and anhydrous calcium chloride.

3. Physical parameters of main reagents

Name	Molecular Weight	Refractive Index	Relative Density	mp /℃	bp /℃	Solubility：g /100ml Solvent		
						Water	Alcohol	Ether
n-Butanol	74.12	1.3993	0.81	−88.9	117.2	7.920	soluble	soluble
1-Bromobutane	137.02	1.4398	1.299	−112.4	101.6	insoluble	soluble	soluble

Experimental Procedure

1. Bromine substitution Put 2~3 boiling chips and 10ml water into a 100ml round-bottom flask. Slowly add 12ml (0.22mol) concentrated sulfuric acid. Mix and cool down the mixture to room temperature. Add 7.5ml (0.08mol) *n*-butanol and 10g (0.10mol) sodium bromide and shake the flask. Install reflux condenser with a hydrogen bromide gas trap on the top. Use 5% sodium hydroxide aqueous as absorbent. Be careful not to immerse the funnel into water to avoid suction. The apparatus is shown in Figure 3-1.

Reflux the mixture gently for 40 min. Shake the flask frequently during the reaction. Turn off the heat source and cool the mixture down. Remove the condenser with the gas trap. Set up a simple distillation apparatus. Distill the mixture until the distillate appears to be clear.

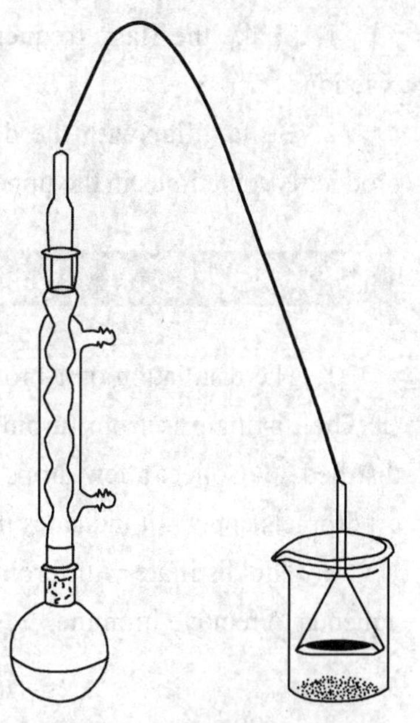

Figure 3-1 Synthesis Apparatus of 1-Bromobutane

图 3-1 1- 溴丁烷的制备装置

2. Purification Transfer the distillate to a separatory funnel, add 10ml water into it and wash the mixture. Allow the funnel to stand until there is a sharp demarcation line between the two layers. Remove the organic layer and save it into another dry separatory funnel. Wash it with 5ml cold concentrated sulfuric acid. Let the mixture stand for a few minutes until the two layers are completely separated. Save the organic layer. Wash the organic layer with 10ml water again, then with 10ml saturated sodium bicarbonate aqueous and finally with 10ml water. Dry the product with a little anhydrous calcium chloride in a dry conical bottle until the liquid is clear. Place the bottle for 0.5h. Decant the clear liquid into a dry distillation flask, distill and collect the distillate between 99~103℃. Weigh the product and calculate the yield.

3. Determine the refractive index (折光率) of 1-bromobutane The yield is about 6~7g (about 52%). Pure 1-bromobutane is a colorless liquid (bp 101.6℃, refractiveindex 1.4399, density 1.276.).

Preview Guide

Preview Requirements

1. Learn the principle and method of preparing 1-bromobutane.

2. Learn reflux operations with hazardous gas absorption device.

3. Review the usage and maintenance of separating funnel and be familiar with purification operation.

Notes

1. Shake the flask frequently during feeding and refluxing process to promote full mixing and reaction.

2. Be familiar with the densities of the reagents. Make sure which layer is the produce as the product is sometimes in the upper layer, while sometimes in the bottom layer.

Experimental Explanation

1. The distillation of 1-bromobutane that is finished can be judged from the following three aspects. ① The distillate is from turbidity (浑浊) to clarification. ② The upper oil layer in the flask has been distilled. ③ Collect a few drops of distillate into a test tube, add a little water and shake the mixture. If no oil droplets appear, it indicates that the product has all been distilled.

2. The distillate with bromine is red after water washing. Add 10~15ml saturated sodium bisulfite aqueous to remove bromine.

$$2NaBr + 3H_2SO_4 \longrightarrow Br_2 + SO_2 + 2H_2O + 2NaHSO_4$$

$$Br_2 + 3NaHSO_3 \longrightarrow 2NaBr + NaHSO_4 + 2SO_2 + H_2O$$

3. Concentrated sulfuric acid can remove a small amount of unreacted *n*-butanol and by-product butyl ether in crude product. Otherwise, *n*-butanol and butyl bromide may form azeotrope (共沸混合物) (bp 98.6℃, containing 13% *n*-butanol), which is difficult to be removed.

Questions

1. Sulfuric acid plays several roles in the reaction. Please summarize.

2. What impurities are contained in the crude products after the reaction? Please explain how to remove them.

3. Why do we wash the organic layer with water before using saturated sodium bicarbonate aqueous to remove the acid?

实验十 正溴丁烷的制备

实验目的

1. 理解以溴化钠、浓硫酸和正丁醇制备正溴丁烷的原理和方法。
2. 掌握带有有害气体吸收装置的回流操作。

实验原理

正溴丁烷是由正丁醇与溴化钠、浓硫酸共热而制得的。

$$NaBr+H_2SO_4 \longrightarrow HBr+NaHSO_4$$

$$n\text{-}C_4H_9OH+HBr \xrightleftharpoons{H_2SO_4} n\text{-}C_4H_9Br+H_2O$$

可能发生的副反应有：

$$n\text{-}C_4H_9OH \xrightarrow[\triangle]{H_2SO_4} CH_3CH_2CH{=}CH_2+H_2O$$

$$2n\text{-}C_4H_9OH \xrightarrow[\triangle]{H_2SO_4} (n\text{-}C_4H_9)_2O+H_2O$$

$$2HBr+H_2SO_4 \xrightarrow[\triangle]{} Br_2+SO_2+2H_2O$$

仪器与试剂

1. 仪器 100ml 圆底烧瓶、球形冷凝管、气体吸收装置、蒸馏装置、分液漏斗、锥形瓶、电炉。

2. 试剂 正丁醇、浓硫酸、溴化钠、氢氧化钠（5%）、饱和碳酸氢钠、无水氯化钙。

实验步骤

1. 溴代 在 100ml 的圆底烧瓶中，加入 10ml 水，慢慢地加入 12ml（0.22mol）浓硫酸，混合均匀并冷却至室温。加入正丁醇 7.5ml（0.08mol），混合后加入 10g（0.10mol）研细的溴化钠，充分振摇，再加入 2~3 粒沸石，装上回流冷凝管，在冷凝管上端接一吸收溴化氢气体的装置，用 5% 的氢氧化钠溶液作吸收剂。注意：切勿将漏斗全部浸入水中，以免倒吸，装置见图 3-1。用小火加热回流约 40min (在此过程中，要经常摇动)。冷却后，改作蒸馏装置，加热蒸出所有溴丁烷。

2. 纯化 将馏出液小心地转入分液漏斗中，用 10ml 水洗涤，小心地将粗产品转入另一干燥的分液漏斗中，用 5ml 浓硫酸洗涤。分去硫酸层，有机层依次分别用水，饱和碳酸氢钠溶液和水各 10ml 洗涤。产物移入干燥的小锥形瓶中，加入少量无水氯化钙干燥，间歇摇动，直至液体透明，时间约 0.5h。将干燥后的产物小心地转入蒸馏烧瓶中。在石棉网上加热蒸馏，收集 99~103℃ 的馏分，称重，计算产率。

3. 测定正溴丁烷的折光率 产量为 6~7g (产率约为52%)。纯的正溴丁烷是无色液体（沸点 101.6℃，相对密度 1.276，折光率 1.4399）

预习指导

预习要求

1. 学习正溴丁烷的制备原理及方法。
2. 学会带有有害气体吸收装置的回流操作。
3. 复习分液漏斗的使用和维护，熟悉洗涤、纯化操作。

注意事项

1. 加料过程中及反应回流时振摇反应瓶，促进充分混合和反应。
2. 熟悉各试剂密度，洗涤时要弄清楚产品何时在上层，何时在下层，不要弄错。

实验说明

1. 正溴丁烷是否蒸完，可从下面三个方面判断：①馏出液是否由浑浊变为澄清；②蒸馏烧瓶中上层油层是否已蒸完；③取一支试管收集几滴馏出液，加入少量水摇动，无油珠出现，则表示有机物已被蒸完。

2. 用水洗涤后馏出液如有红色，是因为含有溴的缘故，可以加入 10~15ml 饱和亚硫酸氢钠溶液洗涤除去。

$$2HBr + H_2SO_4 \xrightarrow[\triangle]{} Br_2 + SO_2 + 2H_2O$$

$$Br_2 + NaHSO_3 \longrightarrow NaBr + NaHSO_4 + SO_2$$

3. 浓硫酸可洗去粗产品中少量的未反应的正丁醇和副产物丁醚等杂质。否则正丁醇和溴丁烷可形成共沸物（bp 98.6℃，含 13% 正丁醇）而难以除去。

思考题

1. 本实验中浓硫酸起何作用？
2. 反应后的粗产物中含有哪些杂质？它们是如何被除去的？
3. 为什么用饱和碳酸氢钠水溶液除酸前，要先用水洗涤？

11 Preparation of Anhydrous and Absolute Ethanol

Experimental Purpose

1. Understand the principle of preparing anhydrous and absolute ethanol from 95% the industrial ethanol.

2. Master reflux and distillation with drying tube.

Experimental Principle

The purity of industrial ethanol is about 95%. To improve the purity, we can reflux industrial ethanol with calcium oxide (quicklime). The water in raw ethanol reacts with calcium oxide to form calcium hydroxide and is removed. Thus, the 99.5% anhydrous (无水) ethanol can be obtained.

$$CH_3CH_2OH + H_2O + CaO \xrightarrow{\text{reflux}} Ca(OH)_2 + CH_3CH_2OH$$

If the anhydrous ethanol reacts with magnesium power further, and the purity can reach 99.95%~99.99%.

$$Mg + 2C_2H_5OH \longrightarrow Mg(OC_2H_5)_2 + H_2$$

$$Mg(OC_2H_5)_2 + 2H_2O \longrightarrow 2C_2H_5OH + Mg(OH)_2$$

Instruments and Reagents

1. Instruments 100ml round-bottom flask, spherical condenser, drying tube, distillation apparatus, electric heating jacket, conical flask and measuring cylinder.

2. Reagents industrial ethanol, calcium oxide, anhydrous calcium chloride, magnesium powder, iodine particles and diethyl phthalate (邻苯二甲酸二乙酯).

Experimental Procedure

1. Preparation of anhydrous ethanol Put 30ml industrial ethanol and 4.3g calcium oxide into a 100ml round-bottom flask. Install the reflux condenser with a calcium chloride drying tube on the top. Heat the mixture by electric heating jacket and reflux it for 30min. After cooling the system a little, modify the reflux unit into a distillation unit with a calcium chloride drying tube connected to the adapter. Distill all the ethanol at the rate of 1~2 drops per second. Measure the weight or volume of the product and calculate the recovery rate.

2. Preparation of the absolute ethanol (绝对乙醇) Put 0.2g magnesium powder and 5ml anhydrous ethanol above into a 100ml round-bottom flask. Set up reflux apparatus with a calcium chloride drying tube on the top. After slightly heating the flask, remove the heat source and put in a few iodine particles immediately (Don't shake the flask). Soon the reaction takes place around the iodine particles, slowly expands, and finally reaches an intense degree. After the reaction, add 20ml anhydrous ethanol above and two boiling chips, and then reflux the mixture for one hour. Then add 1ml diethyl phthalate and reflux the system for 10min. After cooling the system a little, modify the reflux unit into a distillation unit with a calcium chloride (氯化钙) drying tube connected to the adapter. Collect the distillate below 80°C. Measure the weight or volume of absolute ethanol and calculate the recovery rate.

3. Determine the refractive index of absolute ethanol Parameters of absolute ethanol: bp 78.3°C, density 0.7894 and n_D^{20} 1.3614.

Preview Guide

1. Learn the principle of preparing absolute ethanol from 95% the industrial ethanol.
2. Review distillation and reflux operations.

1. Anhydrous and absolute ethanol absorb water violently, so all the instruments in this experiment need to be dried.

2. Iodine can accelerate the reaction. If the reaction has not started after adding iodine particles, heat the flask to promote the reaction.

3. Cotton cannot be filled too tight in calcium chloride drying tube, otherwise the airflow obstruction may cause explosion.

1. Calcium oxide reacts with water to form calcium hydroxide which does not decompose when heated and can be left in the flask in the distillation process.

2. The water in anhydrous ethanol for preparing absolute ethanol cannot exceed 0.5%, otherwise the reaction doesn't work.

1. What would you pay attention to when preparing flammable organic reagents?
2. What would you pay attention to when preparing anhydrous reagents?

实验十一　无水和绝对乙醇的制备

实验目的

1. 理解用 95% 的工业乙醇制备无水乙醇和绝对乙醇的原理。
2. 掌握回流、蒸馏及无水操作。

实验原理

一般工业乙醇的纯度大约为 95%，如果需要纯度更高的无水乙醇，可在实验室里将工业乙醇与氧化钙（生石灰）一起加热回流，使乙醇中的水与氧化钙作用，生成氢氧化钙来除掉水分。这样可得纯度达 99.5% 的无水乙醇，反应式为：

$$CH_3CH_2OH + H_2O + CaO \xrightarrow{\text{reflux}} Ca(OH)_2 + CH_3CH_2OH$$

用氧化钙处理所得的乙醇，如再进一步用金属镁处理，乙醇含量可达 99.95%~99.99%。即绝对乙醇。

$$Mg + 2C_2H_5OH \longrightarrow Mg(OC_2H_5)_2 + H_2$$

$$Mg(OC_2H_5)_2 + 2H_2O \longrightarrow 2C_2H_5OH + Mg(OH)_2$$

仪器与试剂

1. 仪器　圆底烧瓶、球形冷凝管、干燥管、蒸馏装置、电热套、锥形瓶、量筒。

2. 试剂　工业乙醇、氧化钙、无水氯化钙、镁粉、碘粒、邻苯二甲酸二乙酯。

实验步骤

1. 无水乙醇的制备　在 100ml 圆底烧瓶中，加入工业乙醇 30ml 和 4.3g 氧化钙，装上回流冷凝管，在冷凝管的上端安装一个氯化钙干燥管，电热套加热回流 30min。稍冷却后取下冷凝管，改成蒸馏装置，并在接收管的支管上接一氯化钙干燥管，使与大气相通。加热蒸馏直至无乙醇蒸出为止。蒸馏时速度不宜过快，以 1~2 滴/秒的速度为宜。称量无水乙醇的重量或量其体积，计算回收率。

2. 绝对乙醇的制备　装好回流反应装置（顶端附氯化钙干燥管），在 100ml 圆底烧瓶中，加入 0.2g 干燥的镁粉和 2ml 第一步得到的乙醇。在电热套上微热后，移去热源，立即投入几小粒碘粒（注意此时不要振摇），不久碘粒周围即发生反应，慢慢扩大，最后可达比较激烈的程度。反

应结束后，加入 20ml 第一步得到的乙醇，加入两粒沸石，加热回流 1h。然后加入 1ml 邻苯二甲酸二乙酯，再回流 10min，稍冷后，取下冷凝管，改装成蒸馏装置进行蒸馏（引接管附氯化钙干燥管），收集 80℃以下馏分。称量绝对乙醇的重量或量其体积，计算回收率。

3. 测定绝对乙醇的折光率 绝对乙醇的沸点为 78.3℃，密度 0.7894，折光率 n_D^{20} 1.3614。

预习指导

预习要求

1. 学习用 95% 的工业乙醇制备绝对乙醇的原理。
2. 复习蒸馏和回流操作。

注意事项

1. 由于无水乙醇和绝对乙醇具有很强的吸水性，本实验中所用仪器均需干燥。
2. 碘粒可加速反应进行，若加碘粒后仍未开始反应，可适当加热，促使反应进行。
3. 装填氯化钙干燥管时，棉花不能塞得过紧，否则气流不通易引起爆炸。

实验说明

1. 由于氧化钙与水作用生成氢氧化钙，在加热时不分解，故可留在瓶中一起蒸馏。
2. 制备绝对乙醇所用的无水乙醇，水分不能超过 0.5%，否则反应相当困难。

思考题

1. 制备易燃的有机试剂时应注意哪些事项？
2. 制备无水试剂时应注意什么？

12 Preparation of Triphenylcarbinol

Experimental Purpose

1. Be familiar with the application of Grignard reagent and the preparation principle of triphenylcarbinol.
2. Master the method of anhydrous operation for organic reactions.
3. Know the application of steam distillation in organic synthesis.

Experimental Principle

Hydrolysis of nucleophilic additional products of Grignard reagents with aldehyde、ketone、ester and other electron-deficient carbonyl compounds under acidic conditions can be used to synthesize various types of alcohols, which is the most important method of preparing alcohols in the laboratory.

Grignard reagent can be prepared by reaction between halohydrocarbon (卤代烃) and magnesium. Because Grignard reagent is very reactive, it is easily decomposed by water, alcohol, acid and other protic compounds, and is easily oxidized by oxygen in the air. Therefore, the preparation and use of Grignard reagent usually use anhydrous ether as the solvent. One reason is that ether is highly volatile, which can expel most of the air by the ether vapor, so as to reduce the Grignard reagent's contact with the air. The other is that ether can form a complex with Grignard reagent to make it stable. If the halohydrocarbon used has very low activity, anhydrous tetrahydrofuran, with similar properties to ether and a higher boiling point, can be used as the solvent, which can increase the reaction temperature.

In this experiment, triphenylcarbinol is prepared by hydrolysis of the addition product from the reaction between ethyl benzoate and phenyl Grignard reagent. The phenyl Grignard reagent can be prepared with bromobenzene and metal magnesium.

$$C_6H_5Br + Mg \xrightarrow[\text{ether}]{\text{anhydrous}} C_6H_5MgBr$$

$$C_6H_5\overset{O}{\overset{\|}{C}}-OC_2H_5 + C_6H_5MgBr \xrightarrow[\text{ether}]{\text{anhydrous}} (C_6H_5)_2C(OC_2H_5)(OMgBr) \longrightarrow (C_6H_5)_2C{=}O + C_2H_5OMgBr$$

$$(C_6H_5)_2C{=}O + C_6H_5MgBr \xrightarrow[\text{ether}]{\text{anhydrous}} (C_6H_5)_3C-OMgBr \xrightarrow{NH_4Cl/H_2O} (C_6H_5)_3C-OH$$

Instruments and Reagents

1. Instruments a three-necked flask, a dropping funnel, a reflux condenser, a drying tube and a

mercury sealed (汞封的) stirrer, as shown in Figure 3-2.

2. Reagents magnesium, iodine, bromobenzene, anhydrous ether, ethyl benzoate, ammonium chloride, ethanol.

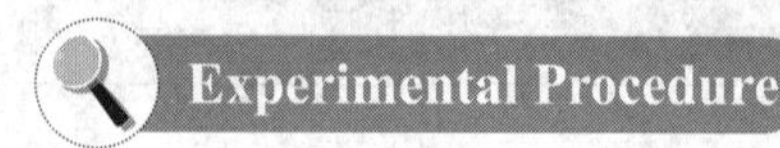

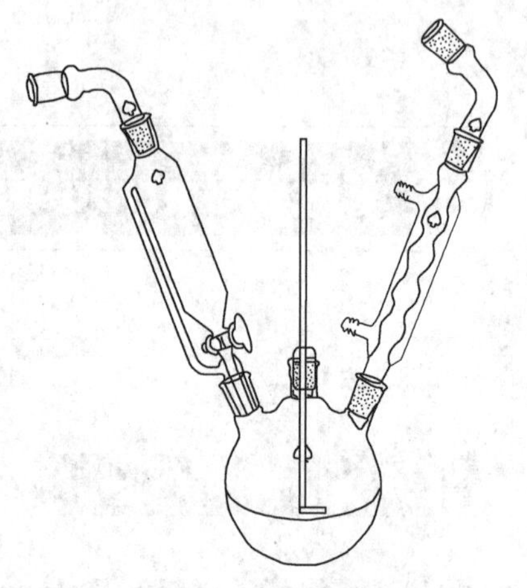

Figure 3-2 The Reaction Device
图 3-2 反应装置图

1. The preparation of phenyl magnesium bromide (苯基溴化镁) Add 0.75g magnesium chips and a small crystal of iodine(碘)into a 100ml three-necked flask which is fitted with a dropping funnel a reflux condenser and a mercury sealed stirrer. Drying tubes are used to prevent the entrance of moisture through the condenser and the dropping funnel during the addition. The reaction device is prepared as shown in Figure 3-2.

Mix 5g bromobenzene(溴苯) and 16ml anhydrous ether in the dropping funnel. Introduce about 8ml of the mixture to the bottom of the flask. Warm the bottom of the flask with your palm. It begins to bubble within a few minutes on the surface of the magnesium chips, indicating the initiation of the reaction. If not, you can heat the flask in a warm water bath to activate the reaction. When the reaction gets intense, turn on the agitator. And the rest of bromobenzen–anhydrous ether solution is added at such a rate that the mixture refluxes gently. After the addition is completed, the flask is heated in a water bath (about 40℃) to keep the mixture refluxing for about 30 minutes to dissolve all of the magnesium chips.

2. The preparation of triphenylcarbinol (三苯甲醇) After 30 minutes (when much of the magnesium chips is gone), remove the heat and cool the flask, first slowly to room temperature, and then in an ice bath. Place 1.9ml of ethyl benzoate(苯甲酸乙酯) and 7ml of anhydrous ether in the dropping funnel and add this solution slowly drop-by-drop with swirling to the Grignard solution in the flask. After the addition is completed, the mixture is refluxed for an hour in a warm water bath. Then, cool the flask in an ice bath and 15ml of saturated ammonium chloride solution is added slowly drop-by-drop to the mixture with constant stirring. Continue to stir for a few minutes. Afterwards, add 2~3 boiling chips to the flask and set up for a simple distillation. The ether is removed by distillation in a warm water bath, and the remaining solution or solid mass is steam-distilled to remove biphenyl (联苯) and unchanged bromobenzene. When the distillate contains no yellow oil beads, the reaction flask is cooled first to room temperature and then in an ice bath. The solid product left in the flask is collected by suction filtration. The crude triphenylcarbinol is weighed and recrystallized from 80% ethanol.

The pure triphenylcarbinol is colorless crystal with boiling point of 380℃, melting point of 160~163℃ and refractive index of 1.1994.

This experiment takes about 8 hours.

Preview Guide

Preview Requirements

1. Understand properties and application of Grignard reagent in organic synthesis.

2. Learn the preparation of Grignard reagent.

3. Learn the anhydrous operation of organic reactions.

4. Review the installation, application and precautions of steam distillation.

Notes

1. Grignard reaction must be carried out under strict moistureless conditions. Therefore, all the equipments and chemicals must be dried. Meanwhile, pay attention to the sealing of the glassware to protect them from any water that may be contained in the air.

2. The solvent used in the experiment is ether, which is flammable, explosive and volatile. Thus, open fire is strictly prohibited during the whole experiment.

3. As the ether is extremely volatile, adequate supplement during the reaction is necessary.

4. Moderate the whole reaction to prevent ether from rushing out from the top of the condensing tube.

5. The reaction mixture will appear pink and may be cured during the preparation of triphenylcarbinol, which is a normal phenomenon.

6. Pay attention to the distillation speed of steam distillation and the liquid level in safety tube.

Experimental Explanation

1. The magnesium metal need to be scraped to remove surface oxides and cut into chips of 2~3mm before use. In order to avoid oxidation, do not touch the magnesium metal.

2. Appropriate (a quarter of a mung bean) iodine can trigger this reaction, while excess will result in more by-products which is adverse to the separation and purification.

3. Make sure the bromobenzen-anhydrous ether solution is added slowly in twice. Otherwise, it will result in more biphenyl byproducts.

4. If there is still a little undissolved solid after the addition of saturated ammonium chloride solution, some dilute hydrochloric acid can be added to facilitate the dissolution.

5. The yellow oil beads in the distillate, which are actually unreacted bromobenzene, should be evaporated entirely so as not to cause problems with purification.

Questions

1. Why do we use saturated ammonium chloride solution for the hydrolysis of the product? Is there any alternative?

2. In this experiment, what if bromobenzene is dropped at a fast rate or all at once?

3. What equipment are used in the preparation of triphenylcarbinol? What are the key points of operation?

4. What should be paid attention in a reaction with Grignard reagent?

实验十二 三苯甲醇的制备

实验目的

1. 熟悉 Grignard 试剂的应用及三苯甲醇制备的原理。
2. 掌握有机反应无水操作的方法。
3. 了解水蒸气蒸馏在合成中的应用。

实验原理

Grignard（格氏）试剂与醛、酮、酯等羰基化合物的亲核加成产物经酸性水解可以得到各种不同类型的醇，是实验室制备醇的最重要的方法。

Grignard 试剂可以通过卤代烃与金属镁反应来制备。由于 Grignard 试剂很活泼，极易被水、醇、酸等质子性化合物分解，同时易被空气中的氧气氧化，因此，Grignard 试剂的制备和使用通常以无水乙醚为溶剂。一是因为乙醚挥发性大，可借乙醚蒸气排开大部分空气，以减少 Grignard 试剂与空气的接触；二是乙醚可以与 Grignard 试剂形成络合物使其稳定。如果所用卤代烃活性甚低，可用无水四氢呋喃作溶剂，它与乙醚性质类似而沸点较高，可借以提高反应温度。

本实验以苯甲酸乙酯为原料，通过与苯基格氏试剂加成再水解制备三苯甲醇，实验中用到的苯基格氏试剂由溴苯与金属镁作用生成。

$$C_6H_5Br + Mg \xrightarrow[\text{ether}]{\text{anhydrous}} C_6H_5MgBr$$

$$C_6H_5COOC_2H_5 + C_6H_5MgBr \xrightarrow[\text{ether}]{\text{anhydrous}} (C_6H_5)_2C(OC_2H_5)(OMgBr) \longrightarrow (C_6H_5)_2C{=}O + C_2H_5OMgBr$$

$$(C_6H_5)_2C{=}O + C_6H_5MgBr \xrightarrow[\text{ether}]{\text{anhydrous}} (C_6H_5)_3C\text{-}OMgBr \xrightarrow{NH_4Cl/H_2O} (C_6H_5)_3C\text{-}OH$$

仪器与试剂

1. 仪器 三颈烧瓶、回流冷凝管、恒压滴液漏斗、搅拌器、干燥管。

2. 试剂 金属镁、碘、溴苯、无水乙醚、苯甲酸乙酯、氯化铵、乙醇。

实验步骤

1. 苯基溴化镁的制备 在干燥的100ml三颈烧瓶中，加入用砂纸擦去表面氧化膜的金属镁（剪成碎屑）0.75g和一小粒碘，烧瓶上分别安装回流冷凝管、恒压滴液漏斗和搅拌器，在冷凝管及恒压滴液漏斗的上口安装氯化钙干燥器，装置如图3-2所示。

将5g溴苯及16ml无水乙醚混合在恒压滴液漏斗中。将恒压滴液漏斗中的混合液放下约8ml到三颈烧瓶中，用手捂住瓶底温热片刻，镁屑表面有气泡产生，表明反应开始。如不反应，可用温水浴稍稍加热使之反应。待反应较激烈时开动搅拌，并缓缓滴入其余的混合液，滴加的速度以维持反应液微沸并有小量回流为宜。滴完后用40℃温水浴加热回流约半小时，使镁完全溶解。

2. 三苯甲醇的制备 用冷水冷却反应瓶，搅拌下将1.9ml苯甲酸乙酯与7ml无水乙醚混合液通过恒压滴液漏斗加入三颈烧瓶中。滴加完毕后，反应混合物温水浴加热回流约1h，使其反应完全。将反应物改为冰水浴冷却，反应物冷却后，在继续搅拌下，向其中慢慢滴加由4g氯化铵配成的饱和水溶液（约15ml），滴完后继续搅拌数分钟。然后改为简单蒸馏装置，投入2~3粒沸石，温水浴加热蒸除乙醚后，瓶中剩余物冷却后析出大量黄色固体。再改为水蒸气蒸馏装置进行水蒸气蒸馏，直至馏出液中不再含有黄色油珠为止。将反应瓶中的剩余物冷却、抽滤，得到三苯甲醇粗产物，称重。用80%的乙醇作溶剂重结晶。

三苯甲醇纯品为无色晶体，沸点380℃，熔点在160~163℃，折光率1.1994。

本实验约需8h。

预习指导

预习要求

1. 理解三苯甲醇的制备原理。
2. 学习Grignard试剂的制备方法。
3. 学习有机反应无水操作的方法。
4. 复习水蒸气蒸馏装置的安装、使用及注意事项。

注意事项

1. 格氏反应需在绝对无水的条件下进行，所用全部仪器和试剂都必须充分干燥，并避免空气中水汽浸入，同时应注意玻璃仪器的密封，防止与空气中水分接触。

2. 实验所用溶剂为乙醚，易燃、易爆、易挥发，整个实验过程严禁使用明火。

3. 反应过程中，乙醚会挥发较多，需要补充一定量的乙醚。

4. 控制整个体系的反应不可太剧烈，防止乙醚从冷凝管上口冲出。

5. 三苯甲醇的制备过程中，反应液会出现淡红色，并且有可能整个反应液固化，这都是正常现象。

6. 水蒸气蒸馏时注意控制蒸馏速度，并时刻注意安全管内水面位置。

实验说明

1. 镁条必须用砂纸充分打磨，去掉表面氧化物至光亮，并用剪刀剪成 2~3mm 长，整个过程不能直接用手接触镁条，避免再引起氧化。

2. 加碘可引发此反应，但不宜多加，有四分之一粒绿豆大小即可，多加会产生较多副产物给分离纯化带来麻烦。

3. 溴苯溶液滴加速度不能太快，更不能一次加完，否则副产物联苯较多。

4. 滴加完饱和氯化铵溶液后，如仍有少量固体未溶，可加少许稀盐酸使之溶解。

5. 馏出液中黄色油珠是未反应的溴苯，必须蒸除干净，以免给纯化带来麻烦。

思考题

1. 实验中为什么要用饱和氯化铵溶液水解产物，还可用何种试剂代替？

2. 在本实验中若溴苯滴得太快或一次加入有什么不好？

3. 三苯甲醇的制备采用了哪些实验装置，操作的关键是什么？

4. 有格氏试剂参与的反应中，应该注意些什么？

13 Preparation of Acetophenone

Experimental Purpose

1. Understand the principles of Friedel-Crafts reaction (傅-克反应) for synthesizing aromatic ketones (芳香酮).

2. Be familiar with the basic methods of Friedel-Crafts reaction for preparing aromatic ketones.

3. Master the installation and use of electric stirring devices (电动搅拌装置) and gas absorption devices as well as some basic operations such as anhydrous operation and treatment of toxic gases (有毒气体).

Experimental Principle

Friedel-Crafts acylation involves the acylation of aromatic rings. Aromatic ketones are always prepared from aromatics and acylating agents (酰化试剂) with Lewis acid catalysts. Typical acylating agents are acyl chlorides (酰氯) and anhydrides (酸酐). Typical Lewis acid catalysts are anhydrous aluminum chloride (三氯化铝) and zinc chloride (氯化锌). Aromatic rings with electron-withdrawing groups (such as $-NO_2$, -COOH, $-SO_3H$) are generally failed to undergo Friedel-Crafts reaction. All reactions need to be performed under anhydrous conditions.

In this experiment acetic anhydride reacts with benzene in presence of anhydrous aluminum chloride to form acetophenone (苯乙酮):

$$C_6H_6 + (CH_3CO)_2O \xrightarrow{AlCl_3} C_6H_5COCH_3 + CH_3COOH$$

Instruments and Reagents

1. Instruments three-necked round bottom flask, electric stirrer, dropping funnel, reflux condenser, separating funnel, drying tube and electric heating jacket.

2. Reagents anhydrous aluminum chloride, anhydrous benzene, acetic anhydride, concentrated hydrochloric acid, 5% sodium hydroxide aqueous solution and anhydrous magnesium sulfate (硫酸镁).

Experimental Procedure

Equip a 250ml three-necked flask with a reflux condenser, an electric stirrer and a dropping funnel. Install a calcium chloride drying tube on the top of the condenser which is connected to a hydrogen chloride gas absorption device as shown in Figure 3-3.

Add 20g of crushed anhydrous aluminum chloride and 30ml of anhydrous benzene into a three-necked flask. A mixture of 6ml of acetic anhydride (about 6.5g, 0.063mol) and 10ml of anhydrous benzene is added dropwise slowly with stirring in 20 minutes before refluxing for 30 minutes until no gas escapes.

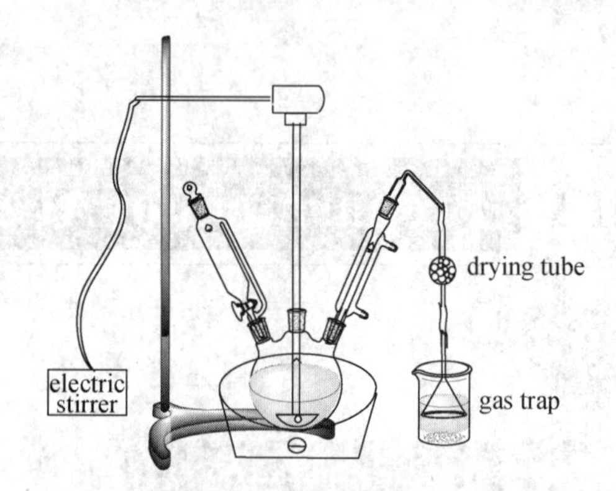

Figure 3-3 The Apparatus For Preparation of Acetophenone

图 3-3 苯乙酮制备装置图

Immerse the three-necked flask in a cold-water bath. Add a mixed solution of 50ml concentrated hydrochloric acid and 50ml ice water slowly with stirring. After the solids are completely dissolved, separate the benzene layer. Wash the combined organic phase successively with 20ml 5% sodium hydroxide solution and 20ml water, dry it using anhydrous magnesium sulfate.

The dried crude product is first distilled off benzene on an electric heating jacket. When the temperature rises to 140℃, stop heating and transfer the crude acetophenone to a clean distillation apparatus and distill using an air condenser. The fraction with boiling point between 198~202℃ is collected to give the product 4~5g (yield: 52%~65%).

Preview Guide

Preview Requirements

1. Review the mechanism and application of Friedel-Crafts acylation reaction.
2. Be familiar with the procedures for preparation of acetophenone.
3. Learn basic operations and precautions such as anhydrous operation and treatment of toxic gases.

Notes

1. The instruments must be absolutely anhydrous, otherwise the reaction will be affected. Wherever the device contacted with air, a drying tube should be installed. This reaction deals with corrosive acids and poisonous fumes. It is advised to be performed in an efficient fume hood, especially during the quenching stage.

2. The quality of anhydrous aluminum trichloride is one of the keys to the success of the experiment. Grind carefully, weigh and feed quickly to avoid prolonged exposure to the air. Weigh them in conical flasks with stoppers and avoid contacting with the skin.

3. As the quenching of the reaction mixture produce enormous amounts of heat, the mixed solution of concentrated hydrochloric acid and ice water should be added dropwise with stirring.

4. Since there are not many final products, a smaller distillation flask should be used. The benzene solution can be added to the distillation flask several times with a separatory funnel.

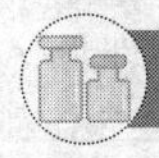

Experimental Explanation

In order to reduce the loss of the product in the experiment, a 2.5cm long glass tube with an outer diameter like the branch tube can be used. The glass tube and the branch tube can be connected by rubber tubes.

Questions

1. Why is benzene required to be thiophene-free? How to remove thiophene from crude benzene?
2. Why should excess benzene and anhydrous aluminum chloride be used in this experiment?
3. What should we pay attention to when using and distilling benzene?

实验十三 苯乙酮的制备

实验目的

1. 理解 Friedel- Crafts 反应（傅 – 克反应）制备芳香酮的原理。
2. 熟悉 Friedel- Crafts 反应（傅 – 克反应）制备芳香酮的基本方法。
3. 掌握机械搅拌装置和气体吸收装置的安装和使用；掌握无水、有毒气体处理等基本操作。

实验原理

芳香酮的制备一般采用傅 - 克酰基化反应，是以酰氯或者酸酐为酰化剂，在路易斯酸催化下进行。常用的路易斯酸催化剂有无水三氯化铝、无水氯化锌等。连有吸电子基团（如 $-NO_2$，$-COOH$，$-SO_3H$ 等）的芳香环一般难以发生傅 – 克酰基化反应。所有傅 – 克反应均需在无水条件下进行。

苯乙酮的制备就是在无水三氯化铝的存在下，苯与酰氯或酸酐作用，芳环上的氢原子被酰基取代制备得到。

仪器与试剂

1. 仪器 三颈烧瓶、电动搅拌、滴液漏斗、冷凝管、干燥装置、HCl 气体吸收装置。

2. 试剂 无水三氯化铝、无水苯、乙酸酐、浓盐酸、5% 氢氧化钠溶液、无水硫酸镁。

实验步骤

在 250ml 干燥三颈瓶中，分别安装搅拌器、滴液漏斗及冷凝管。在冷凝管上端装上氯化钙干燥管，后者再接氯化氢气体吸收装置，如图 3-3。

迅速称取 20g 经研碎的无水三氯化铝，放入三颈瓶中，再加入 30ml 无水苯，在搅拌下滴加 6ml 乙酸酐（约 6.5g，0.063mol）与 10ml 无水苯的混合液（约 20min 滴完）。加完后，在电热套上回流半小时，至无氯化氢气体逸出为止。然后将三颈瓶浸入冷水浴中，在搅拌下慢慢滴入 50ml 浓盐酸与 50ml 冰水的混合液。瓶内固体完全溶解后，分出苯层。水层每次用 15ml 苯萃取两次。合并苯层，依次用 5% 氢氧化钠溶液、水各 20ml 洗涤，苯层用无水硫酸镁干燥。

将干燥后的粗产物先在电热套上蒸出苯，当温度升至 140℃ 时，停止加热，稍冷换用空气冷凝管。收集 198~202℃ 的馏分，产量 4~5g (产率52%~65%)。

预习指导

预习要求

1. 复习 Friedel-Crafts 酰基化反应的机理并了解其应用范围。
2. 熟悉机械搅拌装置和气体吸收装置的安装和使用。
3. 学习无水操作、有毒气体的处理等基本操作及注意事项。

注意事项

1. 仪器必须充分干燥，否则影响反应顺利进行。装置中凡是和空气相接触的地方，应装置干燥管。本实验涉及腐蚀性的酸、有毒气体，因此务必在通风柜中进行。

2. 无水三氯化铝的质量是实验成败的关键之一。研细后称量、投料都要迅速，避免长时间暴露在空气中。为此，可在带塞的锥形瓶中称量。还应避免与皮肤接触，以免被灼伤。

3. 向反应混合物中加入浓盐酸与冰水混合液这一步是放热反应，一定要慢慢滴加，避免氯化氢气体和苯蒸气大量挥发。

4. 由于最终产物不多，宜选用较小的蒸馏瓶，苯溶液可用分液漏斗分数次加入蒸馏瓶中。

实验说明

本实验为减少产品的损失，可用一根 2.5cm 长、外径与支管相仿的玻璃管代替，玻璃管与支管可通过橡皮管连接。

思考题

1. 为什么要求所用的苯不含噻吩？如何除去粗苯中的噻吩？
2. 为什么要使用过量的苯和无水三氯化铝？
3. 使用和蒸馏苯时要注意什么？

PPT

14 Preparation of Adipic Acid

Experimental Purpose

1. Master the operation of electromagnetic stirrer.
2. Learn the principle and method of preparing adipic acid by oxidation of cyclohexanol.
3. Know the application of phase transfer catalyst in heterogeneous organic synthesis.

Experimental Principle

Adipic acid is a kind of common material in the industrial synthesis, which is one of the main raw materials for manufacture of nylon 66. There are a lot of methods for the synthesis of adipic acid, which can be prepared by diethyl malonate or ethyl acetoacetate reacting with the corresponding halogenated hydrocarbon. It is often prepared by oxidizing cyclohexanol with potassium permanganate or nitric acid in the laboratory. The reaction mechanism is that cyclohexanol is oxidized by oxidant to produce cyclohexanone, which is further oxidized to break bonds to produce adipic acid.

$$\text{cyclohexanol (OH)} \xrightarrow[\text{TEBAC, 50°C}]{KMnO_4/Na_2CO_3} \text{cyclohexanone (O)} \xrightarrow{[O]} \text{COONa, COOK} \xrightarrow{H_2SO_4} \text{COOH, COOH}$$

$$\text{cyclohexanol (OH)} \xrightarrow[\text{Room temperature}]{HNO_3} \text{cyclohexanone (O)} \xrightarrow[\text{80°C}]{HNO_3} \text{COOH, COOH} + NO_2\uparrow + H_2O$$

Instruments and Reagents

1. Instruments electromagnetic stirrer, three-necked flask, flask, suction flask, spherical condensing tube.

2. Reagents cyclohexanol, sodium carbonate, potassium permanganate, concentrated sulfuric acid, benzyl triethyl ammonium chloride (BTEAC).

Experimental Procedure

8ml cyclohexanol, 50ml 15% sodium carbonate solution and 1g benzyl triethyl ammonium chloride

(BTEAC) catalyst are added to a 250ml three-necked flask with an electromagnetic stirrer, an agitating magneton and a spherical condensing tube. Meanwhile, fill with condensate water and run the magnetic agitator (the temperature is controlled at 20°C, which is room temperature).

Under the rapid stirring, add 1g of potassium permanganate and then stir for 5 minutes. When the purple color of the reaction solution fades and the brown manganese dioxide appears in the flask, adjust the water temperature to 50°C slowly. Add the remaining potassium permanganate (29g) repeated small in batches (0.5~1g/min) within 30 minutes while keeping stirring. After adding the oxidant, continue stirring at 50°C water bath for 30 minutes. A large amount of manganese dioxide will precipitate in the process of reaction.

Filter by suction the reaction mixture while hot, wash the flask and the residue with mother liquor, and then wash the flask and filter residue with 10ml 10% sodium carbonate solution twice (the total filtrate volume should not exceed 70ml).

Transfer the filtrate to a 250ml beaker. Add 5g of salt powder while stirring (heat to dissolve if necessary). Add 8~10ml concentrated sulfuric acid to the filtrate drop by drop and stir thoroughly until the solution is highly acidic (pH=1~2). After adipic acid has precipitated (write down the amount of concentrated sulfuric acid used), cool it in an ice-water bath for 20 minutes, filter, wash the container with mother liquor and wash the crystals with 5ml ice water. Then filter, dry and obtain rough crystals.

Recrystallization method: Put the rough crystals in cold water at a ratio of 4ml water per 1g, heat to boiling for 3~5 minutes to obtain a transparent solution. Filter under reduced pressure and stand for 30 minutes, then filter to obtain the product and calculate the yield. The yield is about 2~4.5g.

Pure adipic acid is white crystal or crystalline powder. Its boiling point is 337°C, melting point is 152°C and relative density is 1.360.

Preview Guide

Preview Requirements

1. Know the types and principles of commonly used phase transfer catalyst.
2. Be familiar with the application of salting out in organic chemistry.
3. Review the general process of recrystallization.
4. Consult the physical constants of main reagents and raw material used in this experiment.

Notes

1. After the materials are added, continue stirring in water bath to make the reaction complete. But this step must be performed after the reaction temperature no longer rises.

2. Potassium adipate salt is easy to be mixed in manganese dioxide filter residue, so it must be washed with sodium carbonate solution.

3. Due to certain water solubility of adipic acid (1.46g/100ml), the dosage of 10% sodium carbonate for washing should be strictly controlled. If the dosage is too much (total volume > 100ml), the output of the product will decrease significantly. Concentrate the alkaline solution properly (< 70ml), then add a small amount of activated carbon to decolorize. After filtering and cooling, concentrated sulfuric acid is

used to precipitate as much adipic acid as possible.

4. The reaction solution must be thoroughly stirred when the concentrated sulfuric acid is added. Due to the escape of carbon dioxide gas in the acidification process, a small amount of the concentrated sulfuric acid should be added slowly several times. Determination the pH values of solution should be performed after adding acid and stirring for 1~2 minutes.

Experimental Explanation

1. Oxidation of cyclohexanone by potassium permanganate is an exothermic reaction. The addition rate of potassium permanganate must be controlled to prevent the temperature from rising quickly.

2. When the reaction yield of adipic acid is low, adipic acid may be difficult to be separated from the acidic solution. At this time, salting out can be used to reduce the solubility of adipic acid to improve the yield of the product.

3. In addition to potassium permanganate which is used to prepare adipic acid, nitric acid also can be used to prepare adipic acid (method of nitric acid for short). In this method, a device with an addition dropping funnel and a spherical condensing tube connected to a gas absorption apparatus must be set up to prevent nitrogen dioxide from escaping into the laboratory because of the toxic nitrogen dioxide Figure 3-4.

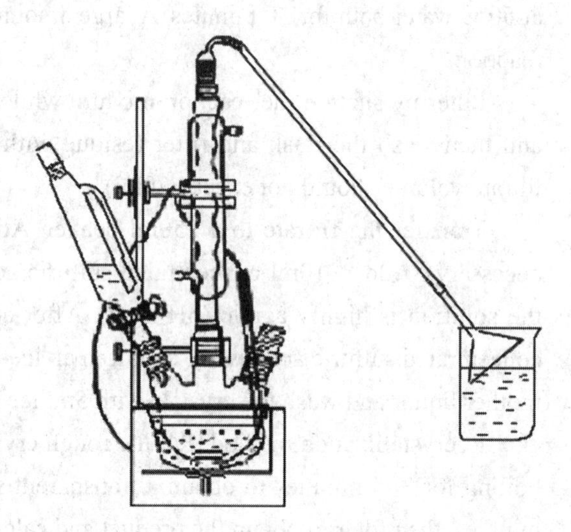

Figure 3-4 Magnetic Stirrer-reflow Devices
图 3-4 磁力搅拌回流装置图

Questions

1. Write the equation of potassium permanganate oxidation cyclohexanol and calculate the theoretical yield of adipic acid according to the chemical equation.

2. Why is it necessary to strictly control the rate of adding potassium permanganate and the temperature of oxidation reaction?

3. What should be noticed when the concentrated sulfuric acid is added into the alkaline solution?

4. What should be pay attention to in the process of filtration, transfer and washing after adipic acid precipitation is completed?

5. Why is it that some of the more aggressive experiments start off with a slower feeding rate and the speed can be accelerated after a while?

实验十四 己二酸的合成

实验目的

1. 掌握电磁搅拌器的操作方法。
2. 学习用环己醇氧化制备己二酸的原理和方法。
3. 了解相转移催化剂在非均相有机合成反应中的应用。

实验原理

己二酸是工业合成常用的原料，如是合成尼龙 66 的主要原料之一。己二酸的制备方法较多，可以采用丙二酸二乙酯、乙酰乙酸乙酯与相应的卤代烃反应制备，实验室常采用高锰酸钾或硝酸氧化环己醇而制得，其反应机理为环己醇被氧化剂氧化生成环己酮，环己酮进一步氧化断键生成己二酸。

$$\text{环己醇(OH)} \xrightarrow[\text{T EBAC, 50°C}]{KMnO_4/Na_2CO_3} \text{环己酮(O)} \xrightarrow{[O]} \text{(COONa)(COOK)} \xrightarrow{H_2SO_4} \text{(COOH)(COOH)}$$

$$\text{环己醇(OH)} \xrightarrow[\text{Room temperature}]{HNO_3} \text{环己酮(O)} \xrightarrow[80°C]{HNO_3} \text{(COOH)(COOH)} + NO_2\uparrow + H_2O$$

实验仪器和试剂

1. 仪器 电磁搅拌器、三口烧瓶、烧瓶、抽滤瓶、球形冷凝管。

2. 试剂 环己醇、碳酸钠、高锰酸钾、浓硫酸、苄基三乙基氯化铵。

实验步骤

在装有电磁搅拌器的 250ml 三口烧瓶中，加入 8ml 环乙醇、50ml 15% 碳酸钠溶液和 1g 苄基三乙基氯化铵催化剂，在三口烧瓶中口装上一支球形冷凝管，通入冷凝水，开动电磁搅拌器（温度控制在 20℃，即室温），在迅速搅拌下，先加入 1g 左右的高锰酸钾，搅拌 5 分钟后，待烧瓶中反应液紫红色消退并有棕色 MnO_2 出现时，调节温度，使水温缓慢升至 50℃；继续搅拌下，分批少量多次 (0.5~1g/min) 地加入剩余的高锰酸钾（29g），严格控制在 30 分钟内加完。氧化剂加完后，在 50℃ 的水浴中继续搅拌 30 分钟。反应过程中，有大量二氧化锰沉淀生成。

趁热将反应物抽滤，用母液洗涤烧瓶和滤渣后，再用10ml 10%的碳酸钠溶液洗涤烧瓶和滤渣二次（滤液总体积不应超过70ml）。

将滤液转移至一个250ml的烧杯中，在搅拌下，加入5g粉状食盐（必要时可加热溶解），量取8~10ml浓硫酸，用吸管慢慢向滤液中滴加浓硫酸，充分搅拌，直至溶液呈强酸性（pH=1~2），已二酸沉淀析出（记下浓硫酸的用量），冰水冷却20分钟后，抽滤，用母液洗涤容器，晶体最后再用5ml冰水洗涤一次，再抽滤，晾干，得粗结晶。

重结晶方法：按照每1g加入4ml水的比例将产品置于冷水中，加热至沸3~5min得到透明溶液，减压抽滤，静置，冷却30min，抽滤即得产品，计算重结晶收率。产量2~4.5g。

纯已二酸是白色结晶或结晶性粉末，沸点337℃，熔点152℃，相对密度1.360。

预习指导

预习要求

1. 了解常用相转移催化剂的类型及作用原理。
2. 熟悉盐析在有机化学中的运用。
3. 复习重结晶的一般过程。
4. 查阅本实验的主要试剂及原料主要物理常数。

注意事项

1. 原料加完后，在水浴中继续搅拌是为使反应进行得完全。但这一步必须在反应温度不再上升后进行。

2. 在二氧化锰滤渣中易夹杂已二酸钾盐，故必须用碳酸钠溶液把其洗涤下来。

3. 已二酸因有一定的水溶性（1.46g/100ml），应严格控制10%碳酸钠洗涤液的用量，若用量过多（总体积>100ml）时，产品产量将显著下降。此时可将盛有碱性滤液的烧杯置于电炉上，适当浓缩后（<70ml）再加入少量活性炭脱色，抽滤冷却后，再用浓硫酸酸化，以促使已二酸沉淀尽量析出。

4. 反应液在滴加浓硫酸酸化时，一定要充分搅拌，因酸化过程有二氧化碳气体逸出，应少量多次缓缓滴加。每次用广泛pH试纸测试溶液pH值，都应在加酸搅拌1~2分钟后进行。

实验说明

1. 高锰酸钾氧化环已酮是放热反应，必须控制好添加高锰酸钾的速度，以免温度上升太快使反应失控。

2. 当产物已二酸产率较低时，已二酸可能难以从已酸化的酸性溶液中析出。此时可采用加入粉状食盐（先加热溶解）盐析的办法，降低已二酸的溶解度以提高产品的收率。

3. 除了本实验使用的高锰酸钾作为氧化剂氧化环已酮制备已二酸之外，硝酸也可以氧化环已酮制备已二酸（简称硝酸法）。在硝酸法中，由于会产生有毒的二氧化氮气体，故需要在回流操作中安装尾气吸收装置，防止二氧化氮逸散在实验室内。如图3-4所示。

思考题

1. 写出高锰酸钾氧化环己醇的反应方程式，根据方程式计算己二酸的理论产量。

2. 做本实验时，为什么必须严格控制添加高锰酸钾的速度和氧化反应的温度？

3. 碱性反应液滴加浓硫酸应注意什么？

4. 己二酸析出完全后，产品在抽滤、转移、洗涤过程中应注意什么？

5. 为什么一些反应比较剧烈的实验在开始时加料的速度要慢？等反应开始一段时间后可以适当地加快加料速度？

15 Preparation of Cinnamic Acid

Experimental Purpose

1. Be familiar with the principle and method of preparing cinnamic acid by Perkin reaction.
2. Master the basic principle and operation method of steam distillation.
3. Consolidate the recrystallization method.

Experimental Principle

Aromatic aldehydes and carboxylic anhydrides are catalyzed by a weak base to form α, β-unsaturated acids, which is known as Perkin reaction. The essence of this reaction is aldol condensation between anhydride and aromatic aldehyde. The catalyst used is generally the potassium salt or sodium salt of the carboxylic acid corresponding to the anhydrides. Potassium carbonate or tertiary amine can also be used as the catalyst. The reaction formula is as follows:

$$C_6H_5CHO + (CH_3CO)_2O \xrightarrow{CH_3COOK} C_6H_5CH{=}CHCOOH + CH_3COOH$$

Instruments and Reagents

1. Instruments 100ml round-bottom flask (A), 500ml round-bottom flask (B), air condenser, 200°C thermometer, electric heating set, drying tube, apparatus for steam distillation, Buchner funnel, suction filter bottle, erlenmeyer flask.

2. Reagents Benzaldehyde, anhydrous potassium acetate powder, acetic anhydride, sodium carbonate solid, concentrated hydrochloric acid, 50%ethanol, activated carbon.

Experimental Procedure

To a dry 100ml round-bottom flask (A) 3.0g of freshly fused and finely grinded anhydrous potassium acetate powder, 5.0ml (0.05mol) of freshly distilled benzaldehyde and 7.0ml of acetic anhydride are added, swirl the flask to mix the reactants well. Install the air condensation pipe, heating and reflowing on the electric heating sleeve. First, heat it to about 160°C for 45 minutes, then raise the temperature to 170~180°C for 1.5 hours. If the experiment needs to stop midway, fit a drying tube containing anhydrous calcium chloride on the top of the condenser to prevent moisture in the air from entering the reaction system, which may affect the experimental results.

Decant the hot reaction mixture into 50ml of water in a 500ml round-bottom flask(B). Rinse the 100ml round-bottom flask (A) twice with 50ml of boiled water and pour the wash solution into the 500ml flask (B). While shaking the flask thoroughly, slowly add a small amount of sodium carbonate solid(about 7~7.5g)to the flask until the mixture becomes weakly basic. Check the basicity of the mixture with litmus test paper until it turns blue. Assemble an apparatus for steam distillation, and then distil the unreacted benzaldehyde from the flask until the distillate is free of oil beads. Then, allow the solution to cool for a while and add a small amount of activated carbon to the flask (B). After heat the solution under reflux for about 10 minutes, filter the hot solution to a 100ml beaker. Allow the filtrate to cool down to room temperature and acidify it by slow addition of concentrated hydrochloric acid with strong stirring until the evolution of carbon dioxide ceases. Cool the solution in a cold-water bath to allow the cinnamic acid to crystallize completely. Collect the product by vacuum filtration, wash it with a small amount of water, drain the water off thoroughly. Dry it below 100℃ (or dry it naturally in air), yielding 7.5~9g. The product can also be purified by recrystallization from hot water or 50% ethanol.

Pure cinnamic acid is a colorless crystal. Melting point 133℃, Relative density 1.046~1.052, Refractive index 1.619~1.623. Time about 8 hours.

Preview Guide

Preview Requirements

1. Review the principle of Perkin reaction to prepare cinnamic acid and its application scope.
2. Learn the basic principle and operation methods of steam distillation.
3. Review the operation methods and precautions of steam distillation and recrystallization.

Notes

1. Benzaldehyde is an irritant, acetic anhydride and hydrochloric acid are corrosive liquids, do not allow them to contact with your skin and avoid breathing the vapor from all these compounds.
2. Do not heat the flask containing benzaldehyde and acetic anhydride directly on a flame.
3. Pay attention to the temperature. If the temperature is too high, the reaction will be fierce, resulting in forming a large amount of resinous substances, reducing the production of cinnamic acid.
4. Heating under reflux to control the reaction to a slightly boiling state. If the reaction solution is boiled violently, it will affect the condensation of acetic anhydride vapor and affect the yield.
5. If drying with an infrared lamp, be careful to control the temperature below 100℃.

Experimental Explanation

1. Heating the crystal potassium acetate in an evaporating dish until it is melted, continue heating and stirring. When the temperature is about 120℃, the solid will appear. Continue to increase the heating power until the potassium acetate melts again. Stop heating, put it in the dryer for cooling, crushing and reserving. If anhydrous potassium carbonate is used in this reaction, its catalytic effect is better than that of anhydrous sodium carbonate.

2. The benzaldehyde used in this experiment cannot contain benzoic acid, because benzaldehyde will be partially oxidized to produce a small amount of benzoic acid after long-term storage. The existence of benzoic acid would influence the reaction and it is difficult to remove benzoic acid from benzaldehyde, so it needs to be purified before use.

3. Aqueous sodium carbonate solution cannot be replaced by sodium hydroxide solution in the reaction. Because the unreacted benzaldehyde may undergo disproportionation reaction in this case, and the benzoic acid generated is difficult to be separated and purified.

Questions

1. What kind of aldehydes can undergo Perkin reaction?
2. Why is steam distillation used in this reaction?
3. What component is removed from the crude product by steam distillation? Can the crude product be purified by other methods instead of steam distillation?

实验十五 肉桂酸的制备

实验目的

1. 熟悉利用 Perkin 反应制备肉桂酸的原理和方法。
2. 掌握水蒸气蒸馏的基本原理及操作方法。
3. 巩固重结晶的操作方法。

实验原理

芳香醛与羧酸酐在弱碱催化下生成 α,β-不饱和酸的反应称为 Perkin 反应。此反应实质是酸酐与芳醛之间的羟醛缩合，所用催化剂一般是酸酐对应的羧酸钾盐或钠盐，也可以使用碳酸钾或叔胺作催化剂。其反应式为：

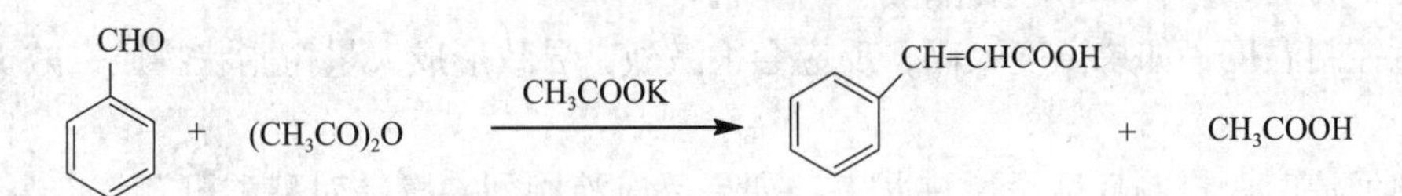

仪器与试剂

1. 仪器 100ml 圆底烧瓶 (A)、500ml 圆底烧瓶 (B)、空气冷凝管、200℃ 温度计、电热套、干燥管、水蒸气蒸馏装置、布氏漏斗、抽滤瓶、锥形瓶。

2. 试剂 苯甲醛、无水乙酸钾粉末、乙酸酐、碳酸钠固体、浓盐酸、50% 乙醇溶液、活性炭。

实验步骤

在干燥的 100ml 圆底烧瓶 (A) 中加入 3g 新鲜熔融并精细研磨的无水乙酸钾粉末，5ml（0.05mol）新蒸馏过的苯甲醛和 7ml 乙酸酐，旋转烧瓶使反应物充分混合。装上空气冷凝管，在电热套上加热回流。先加热至 160℃ 左右，保持 45min，然后升温至 170~180℃，继续反应 1.5h。如果实验需中途停顿，则应在冷凝管上端接一个装有氯化钙的干燥管，以防空气中水分进入反应体系，影响实验结果。

将反应物趁热倒入盛有 50ml 水的 500ml 圆底烧瓶（B）内，100ml 圆底烧瓶（A）用 50ml 沸水分两次洗涤，洗涤液也倒入 500ml 烧瓶（B）中。充分摇动烧瓶，并缓慢加入少量（7~7.5g) 碳酸钠固体，直至反应混合物呈弱碱性，用石蕊试纸检查混合物的碱度，直到试纸变成蓝色。然后进行水蒸气蒸馏，蒸出未反应的苯甲醛直至馏出液无油珠状为止。让溶液冷却一段时间，然后在烧瓶（B）中加入少许活性炭，加热回流 10min，趁热过滤到 100ml 的烧杯中。滤液冷却至室温后，缓慢加入浓盐酸使其酸化，然后快速搅拌直到无二氧化碳产生。将溶液放入冷水浴中，搅拌冷却，使肉桂酸完全结晶。减压过滤，产物用少量水洗净，挤压除去水分，在100℃以下干燥（或在空气中自然干燥），产量7.5~9g。产物也可用热水或50%乙醇重结晶纯化。

纯肉桂醛为无色结晶，熔点为133℃，相对密度为1.046~1.052，折光率为1.619~1.623。本实验需8h。

预习指导

预习要求

1. 复习Perkin反应制备肉桂酸的反应及其应用范围。
2. 学习水蒸气蒸馏的基本原理及操作方法。
3. 复习水蒸气蒸馏和重结晶的操作方法及注意事项。

注意事项

1. 苯甲醛具有刺激性物，醋酸酐和盐酸是腐蚀性液体，主要不要直接接触到皮肤，避免吸入这些化合物的蒸汽。
2. 不要直火加热装有苯甲醛和醋酸酐的烧瓶。
3. 注意控制温度，如果温度太高，反应会很激烈，导致形成大量树脂状物质，减少肉桂酸的生成。
4. 加热回流时，控制反应呈微沸状态，如果反应液剧烈沸腾易对醋酸酐蒸气冷凝产生影响，影响产率。
5. 如用红外灯干燥，应注意控制温度在100℃以下。

实验说明

1. 将乙酸钾晶体置于蒸发皿中加热至熔融，继续加热并不断搅拌。约120℃时出现固体，继续加大火力加热，直到乙酸钾再次熔融，停止加热，置干燥器中放冷，碾碎，备用。另外，无水碳酸钾的催化效果比无水碳酸钠要好。
2. 本实验所用苯甲醛不能含有苯甲酸，因苯甲醛久置会部分氧化产生苯甲酸，不但影响反应的进行，还会混入杂质不易分离，故在使用前需要纯化。
3. 反应中碳酸钠不能用氢氧化钠代替，因为未反应的苯甲醛有可能发生歧化反应，生成的苯甲酸难以分离纯化。

思考题

1. 何种结构的醛能发生Perkin反应？
2. 本反应为什么要采用水蒸气蒸馏？
3. 用水蒸气蒸馏法从粗产品中除去什么成分？粗产品可以用其他方法代替水蒸气蒸馏提纯吗？

16 Preparation of Ethyl Acetate

Experimental Purpose

1. Understand the characteristics and mechanism of esterification (酯化).
2. Master the preparation principle, method and purification technology of ethyl acetate (乙酸乙酯).
3. Consolidate distillation operation and use of separatory funnel.

Experimental Principle

In this experiment, ethyl acetate is prepared from acetic acid and ethanol by heating under the catalysis of concentrated sulfuric acid. The reaction is as follows:

$$CH_3COOH + C_2H_5OH \underset{}{\overset{H_2SO_4}{\rightleftharpoons}} CH_3COOC_2H_5 + H_2O$$

In esterification, alcohols act as nucleophile to attack carboxyl carbonyls. In the presence of protonic acid, carbonyl carbon is more electron deficient, which is conducive to nucleophilic addition of alcohols. If there is no acid, the esterification of acid and alcohol is very difficult. Increasing temperature or using catalysts (催化剂) can speed up esterification reaction and make the reaction reach equilibrium in a short time. Esterification is a reversible (可逆的) reaction. When the reaction reaches equilibrium, the production of ester will not increase any more. In order to improve the production of ester, excessive ethanol can be added, and the volatile characteristics of ethyl acetate can be used. After it is generated, it can be evaporated out of the reaction mixture immediately, and the water of one of the products can be continuously absorbed and removed by dehydrating agent, which will destroy the reversible reaction equilibrium, so that the yield can be improved.

Instruments and Reagents

1. Instruments round bottom flask, reflux condenser, dropping funnel, separating funnel, 100ml three-neck round-bottom flask, filter funnel, distillation head, 200°C thermometer and sleeve, straight condensing tube, water bath, electric heating jacket.

2. Reagents Acetic acid, ethanol, H_2SO_4, NaCl, $CaCl_2$, anhydrous sodium sulfate.

Experimental Procedure

1. Approach One In a 500ml round bottom flask, add 150ml of acetic acid and 23ml of 95% ethanol, drop 7.5ml of concentrated sulfuric acid under shaking, add 2~3 zeolites (沸石), then shake and

mix thoroughly. Install a reflux condensing tube, heat on the water bath and reflux for 30min. When it is a little cold, it will be converted into a distillation unit and distilled in water bath until there is no distillate. Add saturated sodium carbonate solution to the distillate and shake well. The organic phase is alkaline or neutral. The organic phase is separated out with a separatory funnel, and washed once with equal volume of saturated salt water, and then washed with equal volume of saturated calcium chloride solution. The organic phase is separated out and dried with anhydrous sodium sulfate. Filter the dried crude product, put it into a dry distillation flask, add several zeolites and distill it in a water bath. Collect the 73~78℃ fraction, with the output of about 13.1~15.6g.

2. Approach Two Put 10ml of 95% anhydrous ethanol into a 100ml three-neck flask, add 10ml of concentrated sulfuric acid by portion (分批) under shaking. Add 2~3 zeolites, then shake and mix thoroughly. A thermometer and a dropping funnel should be installed respectively on the two sides of the three-neck flask, here the mercury bubble of the thermometer and the end of the dropping funnel should be inserted below the liquid level. In the dropping funnel add 20ml acetic acid and 20ml 95% ethanol. Carefully open the piston of the dropping funnel, firstly add 3~5ml of mixture into the three-neck flask, and close the piston. The middle port of the three neck flask should be equipped with a simple distillation device. After installation, slowly heat it on the electric heating jacket to 110℃, at this time the liquid will steam out. Then, add the left mixture in the dropping funnel to the reaction flask slowly (about 70min). The temperature is controlled at about 120~125℃. After dripping, continue heating for several minutes until the temperature rises to 130~132℃ or no more liquid distills.

Transfer the fraction to the beaker. Add the saturated aqueous $NaHCO_3$ slowly until the pH value is 7 and no bubbles. Transfer the mixture to the separatory funnel. The organic phase (有机相) is separated and washed with 10ml saturated salt water (饱和食盐水). After the lower water layer is separated, the ester layer is washed twice with 10ml saturated calcium chloride. After each static stratification, the lower water layer is discarded.

Transfer the organic phase to a 50ml dry Erlenmeyer flask and dry it with 3g of $MgSO_4$, seal the Erlenmeyer flask with stopper (空心塞) until the solution become clear. Filter it into a dry 60~100ml distillation flask through a funnel, and heat it in a water bath for distillation, collect the fraction between 73~78℃. Weight and calculate the yield.

Ethyl acetate is a colorless transparent liquid, melting point −83.6℃, boiling point 77.1℃, relative density 0.9003, refractive index 1.3723. This experiment takes about 6 hours.

Preview Guide

Preview Requirements

1. Preview the esterification mechanism.
2. Analyze the methods to improve the yield of ethyl acetate.
3. Explore the purification method of ethyl acetate.
4. Learn how to dry liquid organic compounds, the use of desiccants (干燥剂) and precautions.

Notes

1. In this experiment, ethyl acetate is prepared by esterification. The esterification reaction is reversible, so it needs to be kept dry during the experiment.

2. In the preparation of ethyl acetate, the reaction temperature should not be too high, which should be kept between 60~70°C, and the liquid cannot boil. When the temperature is too high, impurities such as ether and sulfite will be produced.

3. The common desiccant include $MgSO_4$, $NaSO_4$, $CaCl_2$, *etc*. The principle of selection of desiccant is not to react with the liquid to be dried, nor to dissolve in the liquid to be dried. Anhydrous calcium chloride is not suitable for drying ethyl acetate because it can form complexes with esters.

Experimental Explanation

1. Because the esterification reaction is reversible, in order to improve the yield, the method of excess alcohol and removal of water in the reaction is generally used. If anhydrous ethanol can be used instead of 95% ethanol, the effect will be better.

2. The purpose of adding saturated sodium carbonate solution is to neutralize and remove acid impurities. The purpose of adding saturated salt water is to remove the remained sodium carbonate. The purpose of adding saturated calcium chloride is to form complex with alcohol to remove alcohol impurities.

Questions

1. Is it appropriate to use excessive acetic acid in this experiment? Why?

2. What are the possible by-products of the experiment? What are the main impurities in the crude ethyl acetate? How to remove them?

实验十六 乙酸乙酯的制备

实验目的

1. 理解酯化反应的特点及其反应机理。
2. 掌握乙酸乙酯的制备原理、方法和纯化过程。
3. 巩固蒸馏及分液漏斗的使用方法。

实验原理

本实验采用乙酸与乙醇为原料，在浓硫酸催化下，加热制得乙酸乙酯，其反应为：

$$CH_3COOH+C_2H_5OH \underset{}{\overset{H_2SO_4}{\rightleftharpoons}} CH_3COOC_2H_5+H_2O$$

在酯化反应中，醇作为亲核试剂对羧基的羰基进行亲核进攻，在质子酸存在时，羰基碳更为缺电子而有利于醇与之发生亲核加成反应。如果没有酸的存在，酸与醇的酯化反应很难进行。升高温度或使用催化剂可加快酯化反应速率，使反应在较短的时间内达到平衡。酯化反应是一个可逆反应，当反应达到平衡后，酯的生成量就不再增加。为了提高酯的产量，可采用加过量的乙醇，并利用乙酸乙酯易挥发的特性，待它生成后立即从反应混合物中蒸出，用脱水剂把生成物之一的水不断吸收除去，破坏此可逆反应平衡，使产量得到提高。

仪器与试剂

1. 仪器 圆底烧杯、回流冷凝管、滴液漏斗、分液漏斗、100ml 三颈圆底烧瓶、布氏漏斗、蒸馏头、200℃ 温度计及套管、直形冷凝管、水浴锅、电热套。

2. 试剂 乙酸、乙醇、浓硫酸、氯化钠、氯化钙、无水硫酸钠。

实验步骤

1. 方法一 在 500ml 圆底烧瓶中，加入 150ml 冰醋酸和 23ml 95% 乙醇，在振摇下滴入 7.5ml 浓硫酸充分摇匀，加 2~3 粒沸石，装上回流冷凝管，在水浴上加热回流 30min。稍冷后，改成蒸馏装置，水浴上蒸馏，蒸至不再有馏出物为止。往馏出液中加饱和碳酸钠溶液，充分摇匀，有机相呈碱性或中性。用分液漏斗分出有机相，有机相加等体积的饱和食盐水洗一次，再用等体积的饱和氯化钙溶液洗一次，分出有机相，用无水硫酸钠干燥。干燥后的粗产品过滤，置于干燥的蒸馏烧瓶中，加几颗沸石，于水浴上蒸馏，收集 73~78℃ 的馏分，产量 13.1~15.6g。

2. 方法二 在 100ml 三颈圆底烧瓶中，放入 10ml 95% 无水乙醇，在振摇下分次加入 10ml 浓硫酸，加完后再充分摇振混匀，加入 2~3 粒沸石，瓶口两侧分别安装温度计和滴液漏斗，温度

计的水银泡和滴液漏斗的尾端均应插到液面以下。将20ml 95%乙醇与20ml冰乙酸混合均匀加入滴液漏斗中。小心开启滴液漏斗的活塞，将3~5ml混合液加入三颈圆底烧瓶中，关闭活塞。在三颈圆底烧瓶的中间口安装简单蒸馏装置，安装完毕后，在电热套上慢慢加热到110℃，这时已有液体蒸出。在此温度下，小心将滴液漏斗的剩余混合物慢慢滴入反应烧瓶中（约70min滴完），控制滴加速度与馏出速度大体相同。滴液初期温度基本稳定在120℃左右，后期会缓缓上升至约125℃。滴完后继续加热数分钟，当温度上升至130~132℃，基本上再无液体馏出时，停止加热。

把收集到的馏出液转移至烧杯中，向馏出液中慢慢加入饱和碳酸钠溶液，振摇混合并用pH试纸检查，直至酯层pH值等于7时，不再有气泡产生。将此混合液转入分液漏斗中充分摇振（注意及时放气），静置分层后分出水层。酯层用10ml饱和食盐水洗涤，在分液漏斗中分离出下面的水层后，酯层再用10ml饱和氯化钙洗两次，每次静置分层后，弃去下面的水层，上面的酯层从分液漏斗上口倒入干燥的50ml锥形瓶中，用大约3g无水硫酸镁干燥，加塞，放置，直到液体澄清。干燥后的液体通过漏斗滤入干燥的60~100ml蒸馏瓶中，用水浴加热进行蒸馏，收集73~78℃的馏分，称重，计算产率。

乙酸乙酯为无色透明液体，熔点–83.6℃，沸点77.1℃，相对密度0.9003，折光率1.3723。本实验约需6小时。

预 习 指 导

预习要求

1. 预习酯化反应机理。
2. 分析提高乙酸乙酯收率的方法。
3. 探讨乙酸乙酯的纯化方法。
4. 学习如何干燥液体有机化合物，使用干燥剂（干燥剂）和预防措施。

注意事项

1. 本实验采用酯化反应制备乙酸乙酯。酯化反应为可逆反应，实验过程中需要保持干燥条件。

2. 制备乙酸乙酯时，反应温度不宜过高，保持在60~70℃，不能使液体沸腾。温度过高时会产生乙醚和亚硫酸等杂质。

3. 常用的干燥剂有$MgSO_4$、$NaSO_4$、$CaCl_2$等，干燥剂的选择原则是不与待干燥液体发生反应，不溶解于待干燥液体中，无水氯化钙不适合干燥乙酸乙酯，因为可与酯形成络合物。

实验说明

1. 因为酯化反应是可逆反应，为了提高产量，一般采用醇过量，以及移除反应中的水的方法。如能用无水乙醇代替质量分数为95%的乙醇效果会更好。

2. 加入饱和碳酸钠溶液目的是为了中和除去酸性杂质，加入饱和食盐水的目的是为了除去残留的碳酸钠，加入饱和氯化钙的目的是与醇形成复合物，以除去醇杂质。

思考题

1. 实验中若采用醋酸过量的做法是否合适？为什么？
2. 实验有哪些可能的副反应？蒸出的粗乙酸乙酯中主要有哪些杂质？如何除去？

17 Preparation of Ethyl Benzoate

Experimental Purpose

1. Master the principle of esterification reaction and the preparation method of ethyl benzoate (苯甲酸乙酯).

2. Learn the use of Dean-Stark trap (分水器) and the purifying methods of organic liquid.

3. Know the principle of azeotropic dehydration (共沸脱水).

Experimental Principle

Carboxylic esters (羧酸酯) are a group of compounds with a wide range of industrial applications. They are often used in flavors or fragrances. Some of them are used as bactericide (杀菌剂) or an esthetics (麻醉剂) because of their fragrance. Carboxylic esters can be prepared by esterification of carboxylic acids and alcohols in the presence of catalysts, or by alcoholysis (醇解) with acylchloride (酰氯), anhydride (酸酐) or nitrile (腈). Esterification is the most important method in laboratory and industry, in the present of sulfuric acid, hydrogen chloride or *p*-toluenesulfonic acid (对甲苯磺酸).

The reaction formula is as follows:

$$C_6H_5COOH + CH_3CH_2OH \underset{}{\overset{H_2SO_4}{\rightleftharpoons}} C_6H_5COOCH_2CH_3 + H_2O$$

Instruments and Reagents

1. Instruments 50ml round-bottom flask, Dean-Stark trap (分水器), spherical condenser，25ml round-bottom flask, distillation head, thermometer with adapter, air condenser, condenser adapter (尾接管), Erlenmeyer flask.

2. Reagents benzoic acid (苯甲酸), absolute ethanol, benzene, sulfuric acid, Na_2CO_3, diethyl ether, water-free magnesium sulfate.

Experimental Procedure

Place 6g benzoic acid, 15ml absolute ethyl alcohol, 12ml benzene, 3ml sulfuric acid and some zeolites into a 50ml round-bottom flask, and mix them together. Install the Dean-Stark trap, add water from the upper end of the Dean-Stark trap to the branch pipe carefully, and then discharge 4.5ml water. Connect the upper end of the Dean-Stark trap with a spherical condenser, as shown in Figure 3-5.

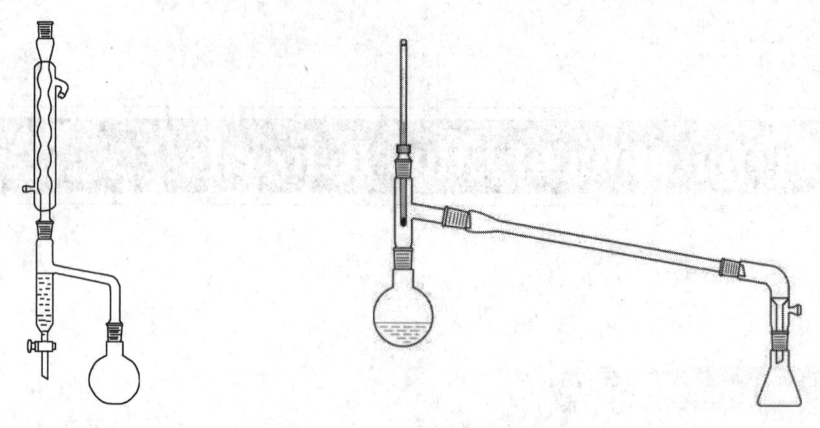

Figure 3-5 Reflux unit with Dean-Stark trap
图 3-5 有分水器的回流装置

Figure 3-6 Distillation unit with air condenser
图 3-6 有空气冷凝器的蒸馏装置

Heat the flask over the water bath to reflux, and keep the return flow tiny at the beginning. And, over time, the upper, middle and lower layers of liquid appear in the Dean-Stark trap. After 2 hours, the middle and lower liquid going to be filled with branches of the Dean-Stark trap, then the water——the middle and lower liquid——is put into the measuring cylinder. The excess ethanol and benzene are going to be filled in branches of the Dean-Stark trap when the distillation continues.

Pour the reaction solution into a beaker with 45ml of water and add the sodium carbonate powder in batches until the solution is neutral (or slightly alkaline) meanwhile there is no carbon dioxide escaping from solution.

Separate the organic layer by a separating funnel. Extract the water phase with 15ml diethyl ether, and then incorporate it into the organic layer. Remove the water from organic layer with water-free magnesium sulfate after that, distill the crude product and remove fractions of diethyl ether at low temperature. Collect the 210~213℃ fractions as product (Figure 3-6) and the yield should be 5g. Ethyl benzoate is a liquid, with b. p. = 213℃, d_4^{20} =1.05, n_D^{20} = 1.504. The experiment takes about 6 hours.

Preview Guide

Preview Requirements

1. Preview the principle of esterification reaction and the work principle of Dean-Stark trap.
2. Preview the principle of azeotropic distillation.
3. Compare the differences of preparation of ethyl benzoate and ethyl acetate in apparatus and operation.

Notes

1. After installing the apparatus (成套仪器), the tightness of all joints should be checked.
2. After installing the Dean-Stark trap, water should added from the upper end of the Dean-Stark trap until arrive at the branch pipe, and then discharge 4.5ml water.
3. When distilling the crude product, the distillation unit should be installed with an air condenser.

Experimental Explanation

1. There are three layers in Dean-Stark trap. The lower layer is the water originally. The fraction from the reaction bottle is ternary azeotrope, which flows into the Dean-Stark trap from the spherical condenser. The ternary azeotrope is divided into two layers. This lower layer of the fraction is the middle layer in the Dean-Stark trap.

2. The purpose of adding sodium carbonate is to remove sulfuric acid and the inactive benzoic acid, so it should be added in batches after grinding, otherwise a lot of foam will be generated and the liquid will overflow.

3. If the crude product contains flocculent (絮状物) which is difficult to stratify, it can be directly extracted with 15ml diethyl ether.

Questions

1. How to improve the yield of ethyl benzoate in this experiment?

2. It is difficult to stratify when there is flocculent or emulsion (乳剂、油状物) between the two phases in the extraction or liquid separation. How to get rid of the flocculent or emulsion when its presence?

实验十七　苯甲酸乙酯的制备

实验目的

1. 掌握酯化反应原理及苯甲酸乙酯的制备方法。
2. 学习分水器的使用及液体有机化合物的精制方法。
3. 了解三元共沸除水原理。

实验原理

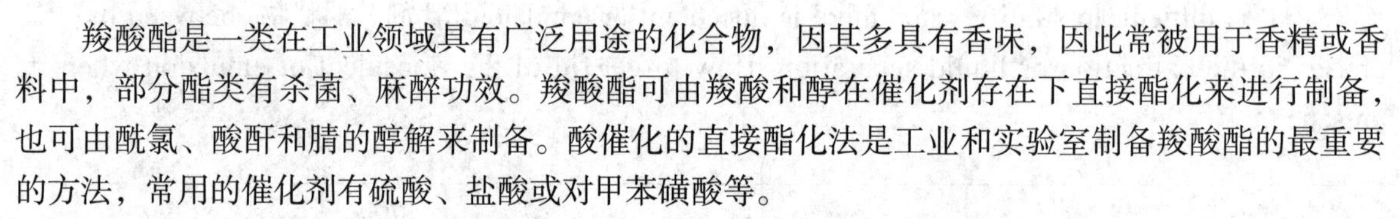

视频

羧酸酯是一类在工业领域具有广泛用途的化合物，因其多具有香味，因此常被用于香精或香料中，部分酯类有杀菌、麻醉功效。羧酸酯可由羧酸和醇在催化剂存在下直接酯化来进行制备，也可由酰氯、酸酐和腈的醇解来制备。酸催化的直接酯化法是工业和实验室制备羧酸酯的最重要的方法，常用的催化剂有硫酸、盐酸或对甲苯磺酸等。

反应式：

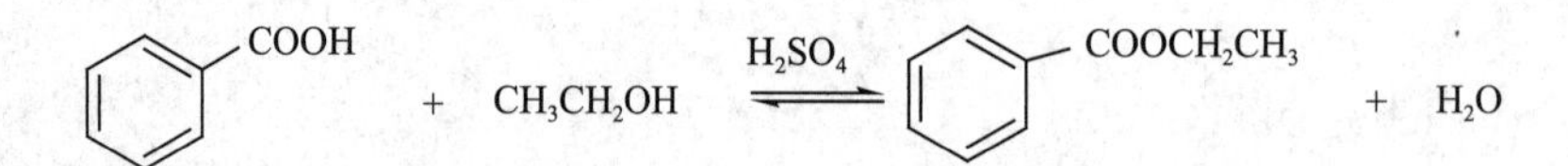

仪器与试剂

1. 仪器　50ml 圆底烧瓶、分水器、球形冷凝管、25ml 圆底烧瓶、蒸馏头、温度计、空气冷凝管、尾接管、锥形瓶。

2. 试剂　苯甲酸、无水乙醇、苯、浓硫酸、Na_2CO_3、乙醚、无水 $MgSO_4$。

实验步骤

在 50ml 圆底烧瓶中加入 6g 苯甲酸、15ml 无水乙醇、12ml 苯和 3ml 浓硫酸，摇匀，加入沸石。在圆底烧瓶口安装分水器，从分水器上端小心加水至分水器支管处，然后再放去 4.5ml 水，分水器上端接球形冷凝管如图 3-5。

将烧瓶在水浴中加热回流，开始时回流速度要慢，随着回流的进行，分水器中出现了上、中、下三层液体。约 2 小时后，中、下层液体接近分水器支管，将中、下层液体放入量筒中。继续蒸馏，以蒸出过量的乙醇和苯。

将反应液倒入盛有 45ml 水的烧杯中，分批加入碳酸钠粉末至溶液呈中性（或弱碱性），无二氧化碳逸出为止。

用分液漏斗分出有机层，将水层溶液用 15ml 乙醚萃取，然后合并至有机层，用无水 $MgSO_4$

干燥有机层。粗产物进行蒸馏，低温蒸出乙醚。收集 210~213℃ 的馏分即为产物图 3-6，产量约 5g。

纯粹苯甲酸乙酯的沸点为 213℃，相对密度 1.05，折射率 1.504。本实验需 6 小时。

预 习 指 导

预习要求

1. 复习酯化反应原理及分水器的工作原理。
2. 复习共沸蒸馏原理。
3. 比较苯甲酸乙酯和乙酸乙酯的制备方法在装置和操作上的差异。

注意事项

1. 安装仪器时，应当检查连接处是否连接紧密。
2. 使用分水器时，应当先加水至支管处，然后放出计算量的水。
3. 产物蒸馏时，应当使用空气冷凝管。

实验说明

1. 分水器中有三层液体，其下层为实验开始前加入的水。由反应瓶中蒸出的馏液为三元共沸物，馏液从冷凝管流入分水器后分为两层，馏液的下层即为分水器中的中层。

2. 加碳酸钠的目的是除去硫酸及未反应的苯甲酸，要研细后分批加入，否则会产生大量泡沫而使液体溢出。

3. 若粗产物中含有絮状物难以分层，则可直接用 15ml 乙醚萃取。

思考题

1. 本实验采用何种措施提高酯的产率？
2. 在萃取和分液时，两相之间有时出现絮状物或乳浊液，难以分层，如何解决？

18 Preparation of Furanmethanol and Furoic Acid

Experimental Purpose

1. Learn about Cannizzaro reaction (康尼查罗反应) and be familiar with the preparation principles and methods of 2-Furanmethanol and 2-Furoic acid.

2. Master the methods of separating and purifying 2-Furanmethanol (2-呋喃甲醇) and 2-Furoic acid (2-呋喃甲酸).

Experimental Principle

Those aldehydes (羰基化合物) without α-H can occur self-redox reaction (自身氧化-还原反应) in strong alkali conditions, one molecule is oxidized to acid, and the other is reduced to alcohol, such reaction is just called Cannizzaro reaction.

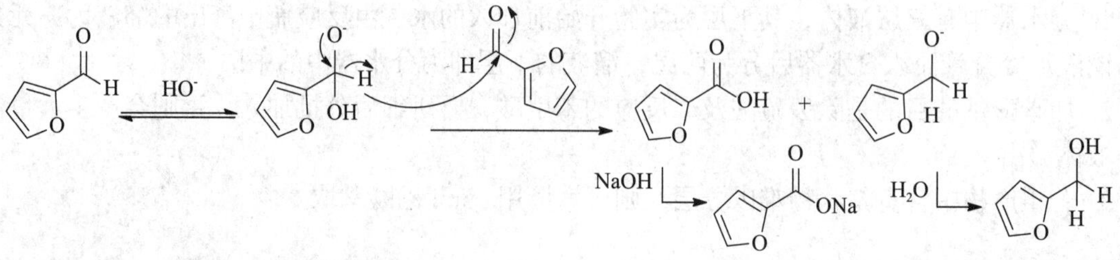

Instruments and Reagents

1. Instruments 250ml beaker, 100ml beaker, drip funnel, separating funnel, Büchner funnel, ordinary funnel, filter paper, Congo red test paper（刚果红试纸）, round bottom flask, distillation head, thermometer of 200℃ and casing, straight condenser tube, air condenser tube, thermostatic water bath, electric heating sleeve.

2. Reagents Furfural (糠醛，又名呋喃甲醛), NaOH, concentrated HCl, diethyl ether, water-free magnesium sulfate (无水硫酸镁).

Experimental Procedure

1. Preparation In a 250ml beaker, add freshly steamed furfural of 19g (about 16.4ml, 0.2mol), immerse the beaker in ice water and cool the reaction system to 5℃, then drip 16ml 33% sodium hydroxide solution slowly using drip funnel, keeping stirring while dripping, using thermometer to test the reaction temperature and strictly control the reaction temperature between 8~12℃. The sodium

hydroxide will be dripped totally for about 20~30min. Then keep stirring at room temperature for about half an hour, you will get a light yellow slurry (浆状物).

2. Separating 2-Furanmethanol Add about 18 ml of water to the reaction mixture to dissolve the precipitation, the solution is dark brown. Pour the solution into separatory funnel and extract with 15ml of ether at one time, totally extract for 4 times using about 60 ml of ether. Combining the ether extract, and remember in this experiment the water layer cannot be discarded! Dry the ether extract with water-less magnesium sulfate or water-less potassium carbonate, filter and steam off the ether using water bath, and then distill 2-Furanmethanol using air condenser tube, collect fractions of 169~172℃. The yield of 2-Furanmethanol is about 7~8g.

Pure 2-Furanmethanol is a colorless or slightly yellow transparent liquid, the boiling point is 171℃, and the density is 1.1269, while its refractive index is 1.4868.

3. Separation of 2-Furoic acid After diethyl ether extraction, the water solution should be stirred and acidified with concentrated HCl to ensure the pH value is 2~3, or you can test it with Congo Red Test Paper till the test paper become blue and does not fade at last. Cooling the solution to ensure 2-Furoic acid completely precipitate, then filtering, using a little cold water to wash the crystal, and then you will get the crude products of 2-Furoic acid.

The rough products should be recrystallized using hot water about 50~100ml, after cooling, the needle-like crystallization of 2-Furoic acid is crystallized, and the product is about 8g.

The melting point of 2-Furoic acid is 133~134℃.

Preview Guide

Preview Requirements

1. Make good preparation by reviewing the theory and seeking information about the chemicals involved in preparation of Furanmethanol and Furoic acid, including Furfural, Furanmethanol and Furoic acid.
2. Explore the special experimental principle of Cannizzaro reaction.
3. Try to compare the use of drip funnel, separatory funnel, Büchner funnel and ordinary funnel.
4. Note special precautions required for separating the ether extract and the water layer.

Notes

1. Furfural will become brown or black if kept for long time, and also will absorb water, therefore, it must be distilled and purified before using.
2. The reaction temperature should be strictly controlled between 8~12℃.
3. After adding sodium hydroxide, keep stirring until the reaction fluid become viscous.
4. Adding too much water will cause product loss.
5. Concentrated HCl should be added enough to ensure the pH value is 2~3, so that 2-Furoic acid can be completely crystallized.

Questions

1. Why freshly steamed Furfural must be used?

2. How to cool the Furfural to 5℃ ? Why the reaction temperature should be strictly controlled between 8~12℃ ?

3. What is the purpose of adding sodium hydroxide?

4. Why concentrated HCl should be added enough, how to do this?

实验十八 呋喃甲醇和呋喃甲酸的制备

实验目的

1. 了解坎尼查罗反应。熟悉呋喃甲醇和呋喃甲酸的制备原理及方法。
2. 掌握分离、纯化呋喃甲醇和呋喃甲酸的方法。

实验原理

视频

不含 α-H 的醛类与浓的强碱溶液作用，可发生自身氧化还原反应，一分子醛被氧化为酸，另一分子醛被还原为醇，此反应称为坎尼查罗反应。

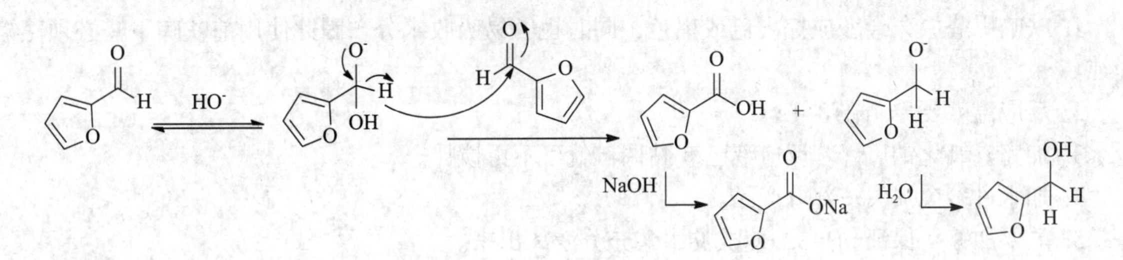

仪器与试剂

1. 仪器 250ml 烧杯、100ml 烧杯、滴液漏斗、分液漏斗、布氏漏斗、普通漏斗、滤纸、圆底烧瓶、蒸馏头、200℃温度计及套管、直形冷凝管、空气冷凝管、水浴锅、电热套。

2. 试剂 呋喃甲醛、氢氧化钠、乙醚、浓盐酸、无水硫酸镁等。

实验步骤

1. 制备 在 250ml 烧杯中，加入新蒸的呋喃甲醛 19g (约16.4ml，0.2mol)，将烧杯浸入冰水中冷却至 5℃左右，用滴液漏斗缓缓滴入 16ml 33% 氢氧化钠溶液，边滴加边持续搅拌，控制滴加速度，使反应体系温度保持在 8~12℃，在 20~30min 之内将氢氧化钠滴加完毕，于室温下静置半小时左右，得到一淡黄色浆状物。

2. 分离呋喃甲醇 向反应混合物中加入约 18ml 的水使沉淀溶解，此时溶液呈暗褐色。将溶液倒入分液漏斗中，每次用 15ml 乙醚萃取 4 次，合并乙醚萃取液，注意本实验中的水层不可弃去。用无水硫酸镁或无水碳酸钾干燥，过滤后先水浴蒸去乙醚，再蒸馏呋喃甲醇，收集 169~172℃ 的馏分。呋喃甲醇产量 7~8g。

纯呋喃甲醇为无色或略带淡黄色的透明液体，沸点 171℃，密度 1.1269，折光率 1.4868。

3. 分离呋喃甲酸 乙醚萃取后的水溶液在搅拌下用浓盐酸酸化至溶液 pH 值达到 2~3，或者以刚果红试纸检验，完全变蓝后不褪色，冷却后即有大量呋喃甲酸晶体析出。抽滤，晶体用少量冷水洗涤，得粗产品呋喃甲酸。

粗产品可以用 50~100ml 热水重结晶，冷却后，得到呋喃甲酸针状晶体，产品重约 8g。

呋喃甲酸的熔点为 133~134℃。

预习指导

预习要求

1. 预习呋喃甲醇和呋喃甲酸制备相关的理论原理。查找实验涉及的化合物信息及理化参数，主要包括呋喃甲醛、呋喃甲醇和呋喃甲酸。
2. 探索本实验特殊的反应原理 Cannizzaro reaction。
3. 试比较滴液漏斗、分液漏斗、布氏漏斗及普通漏斗的用法。
4. 注意分离乙醚萃取层和水层时需要特殊留意的事项。

注意事项

1. 呋喃甲醛放久会变成棕褐色或黑色，同时也容易吸收水分。因此使用前呋喃甲醛必须蒸馏提纯。
2. 反应温度要控制在 8~12℃。
3. 加完氢氧化钠后，若反应液已变黏稠，就可不再搅拌。
4. 加水过多会损失一部分产品。
5. 酸要加够，保证 pH=3，使呋喃甲酸充分游离出来。

思考题

1. 为什么必须使用新蒸的呋喃甲醛？
2. 如何冷却反应原料至 5℃？为什么反应温度必须严格控制在 8~12℃？
3. 加入碳酸钠的目的是什么？
4. 浓盐酸为什么要加足够量，如何保证加够？

19 Preparation of Acetanilide

Experimental Purpose

1. Be familiar with the preparation acetanilide (乙酰苯胺) by the reaction of aniline (苯胺) and glacial acetic acid or acetic anhydride.

2. Master the application of simple fractionation organic synthesis.

Experimental Principle

The acylation (酰化) of aromatic amines (芳香胺) plays an important role in organic synthesis. As a protective measure, primary and secondary aromatic group can be converted into their acetyl derivatives (酰基衍生物) to reduce the sensitivity of aromatic amines to oxidants and prevent them from being destroyed in the synthesis. The acetyl group can be removed from the nitrogen atom by hydrolysis of acetanilide derivatives in either an aqueous acidic or aqueous basic solution. And after the acylation of amino group, the activated ability of amino group in electrophilic substitution reactions is reduced (especially in halogenated reactions (卤代反应), which makes the positioning capability (定位能力) of amino group from stronger activity I positioning group to medium activity ones, the replaced by diversified into unitary substitution. Moreover, due to the space effect of acetyl group, the reaction of electrophilic reagent with acetyl group is often selective, and the resulting products are mostly para-substitutions..

The commonly used acylation reagent in this reaction is acetyl chloride, acetic anhydrides, acetic acid. The reaction, in general, proceeds most rapidly with acetyl chloride, slower with acetic anhydrides, and so slowly with acetic acid themselves that an elevated temperature is required. The following two alternatives have been listed for this experiment.

(1) Used glacial acetic acid as the acylation agent:

$$C_6H_5NH_2 + CH_3COOH \longrightarrow C_6H_5NHCOCH_3 + H_2O$$

(2) Used acetic anhydride as the acylation agent:

$$C_6H_5NH_2 + HCl \longrightarrow C_6H_5NH_3^+Cl^- \xrightarrow[CH_3COONa]{(CH_3CO)_2O} C_6H_5NHCOCH_3 + CH_3COOH + H_2O$$

Instruments and Reagents

1. Instruments 50ml round-bottom flask, 500ml beaker, fractional column, thermometer, Büchner funnel.

2. Reagents aniline, glacial acetic acid, acetic anhydride, sodium acetate (乙酸钠), zinc powder, activated carbon, hydrochloric acid.

Experimental Procedure

1. Method one To a 50ml round-bottom flask add 10.0 ml the freshly distilled aniline, 15ml glacial acetic acid, and 0.1g zinc powder. Assemble the flask with a fractional column (Figure 3-7), and fit a thermometer on the top of the column. Heat slowly the reaction solution under gentle reflux and control the flame to keep the temperature of the distilling vapor being about 100~110℃ for 1.5 hours. The water formed in the reaction could be distilled completely (containing a small amount of acetic acid). When the temperature of the distilling vapor fluctuates, the reaction has ended, then stop heating. Pour the hot solution slowly into a beaker with 200ml of cold water. Cool the aqueous solution with vigorous stirring to make the acetanilide completely precipitate as fine crystals. Collect the crude product in a Büchner funnel by vacuum filtration, and wash the product with 5~10ml of ice-water to remove the remaining acetic acid, and the yield is about 9~10g.

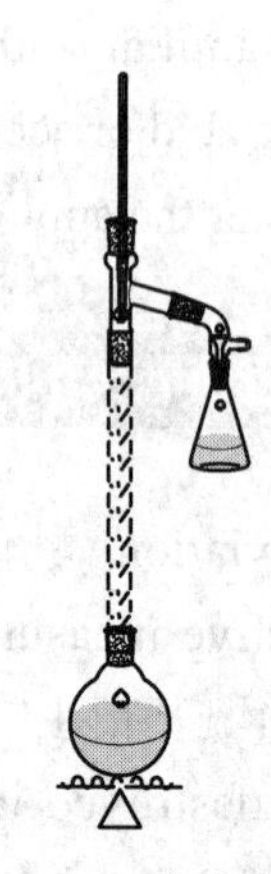

Figure 3-7 The Scheme of Spines Fractionating Column Reflux Reactor
图 3-7 刺形分馏柱回流反应装置

2. Method two In a 500ml beaker, dissolve 5ml concentrated hydrochloric acid in 120ml distilled water, and then add 5.6ml aniline under stirring. After aniline is dissolved, a small amount of activated carbon (about 1g) is added. Boil the solution for 5 minutes and remove the activated carbon and other insoluble impurities while it is hot. Transfer the filtrate to a 500ml conical flask (锥形瓶) and cool it to 50°C. Add 7.3ml acetic anhydride and shake it to dissolve. Meanwhile, add the solution of 9.0g sodium acetate in 20ml water immediately. The above mixture cooled in an ice bath to allow it to crystallize. Collect the crystal upon a Büchner funnel and wash them with a little cold water. Weigh the products after drying, and the yield was about 5~6g. The acetanilide prepared by this method is pure enough, generally which can be directly used in the required synthesis. For further refinement, recrystallization may be performed with pure water.

The pure acetanilide is colorless and shiny lobular crystal. The melting point of acetanilide is 114.3°C.

This experiment takes about 2~3h.

Preview Guide

1. Learn the principle and method of preparing acetanilide by aniline acylation.
2. Review the fractionation unit installation and operation.
3. Review the physical and chemical properties of aniline, acetic acid, acetic anhydride and acetanilide.
4. Review the basic operation of the vacuum filtration and recrystallization.

1. Aniline turns red-brown after stored for a long time. This may influence the quality of product acetanilide. Fresh distilled aniline is preferred in this experiment.
2. Zinc powder is added to prevent the aniline from being oxidized during reaction. Do not add too much, otherwise the water-insoluble zinc hydroxide would appear in the procedure of work-up. The optimal amount of zinc powder in the present experiment is about 0.1g.
3. To prevent the solid product from adhering to the reactor wall and causing the loss of product yield, it is best to immediately pour the reactant into cold water while it is hot and stir it to remove the excess acetic acid and the unreacted aniline.
4. The activated carbon should not be added to the boiling solution, otherwise it will cause the reaction to boil violently.

1. Why should the temperature of nozzle at the top of the fractionation column be controlled between 100~110°C during the reaction? What happens when the temperature is too high?
2. What method can be used to improve the yield of the product when using glacial acetic acid as acylation reagent to prepare acetanilide?
3. What is the purpose of adding hydrochloric acid and sodium acetate to the acetylation of acetic anhydride?
4. A small amount of oil beads usually appears in the beaker when preparing the saturated solution of acetanilide by recrystallization. Try to explain the reason and give the best solution to deal with it.

实验十九 乙酰苯胺的制备

实验目的

1. 熟悉乙酰苯胺制备的方法及相关反应机理。
2. 掌握简单分馏在合成中的应用。

实验原理

视频

芳胺的酰化产物在有机合成中有着重要的作用。作为一种保护措施，一级和二级芳胺在合成中可被转化为它们的酰基衍生物，以降低芳胺对氧化剂的敏感性，使其不被反应试剂破坏；反应结束后，酰基衍生物可在酸碱催化下水解恢复氨基。同时，氨基经酰化后降低了氨基在亲电取代反应（特别是卤化）中的活化能力，使其由活性很强的第Ⅰ类定位基，变为中等活性，反应由多元取代变为一元取代；同时由于乙酰基的空间效应，亲电试剂与之反应往往具有选择性，生成的产物多为对位取代。

常用的酰化试剂有乙酸、乙酸酐、乙酰氯等。乙酸价廉，但反应活性较低；乙酸酐、乙酰氯反应活性较高，产物较纯，但乙酰氯价格较贵，实验中常用乙酸酐做酰化试剂。

以冰醋酸、乙酸酐作为酰化试剂的反应如下。

（1）用冰醋酸为酰化试剂：

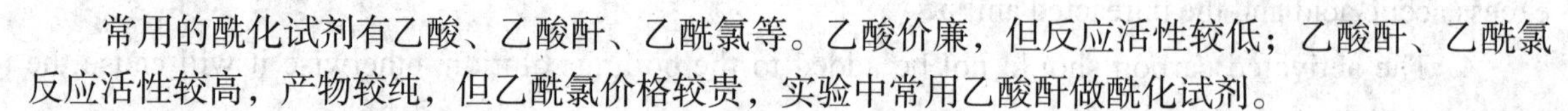

（2）用乙酸酐为酰化试剂：

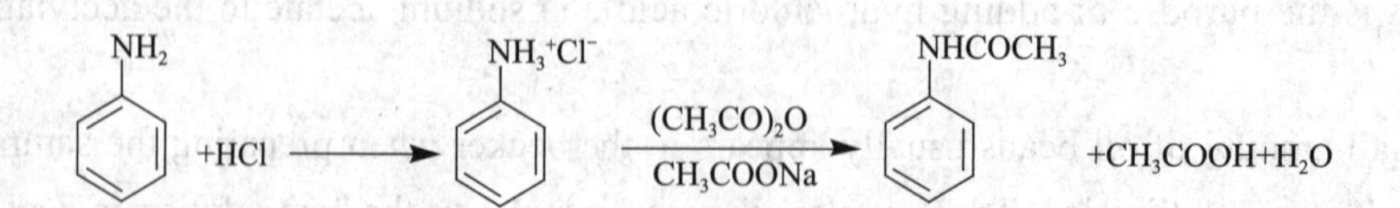

仪器和试剂

1. 仪器 50ml 圆底烧瓶、500ml 烧杯、分馏柱、温度计、布氏漏斗。

2. 试剂 苯胺、冰醋酸、乙酸酐、乙酸钠、锌粉、活性炭、盐酸。

实验步骤

1. 方法一 在 50ml 圆底烧瓶中，加入 10ml 新蒸苯胺、15ml 冰醋酸及少许锌粉（约 0.1g），装上一短的刺形分馏柱（图 3-7），其上端装一温度计，支管通过支管接引管与接收瓶相连，接收瓶外部用冷水浴冷却。

将圆底烧瓶在石棉网上用小火加热，使反应物保持微沸约 15min。然后逐渐升高温度，当温度计示数达到 100℃ 左右时，支管即有液体流出。维持温度在 100~110℃ 之间反应约1.5h，生成的水及大部分冰醋酸已被蒸出，此时温度计示数下降，表示反应已经完成。在搅拌下趁热将反应物倒入 200ml 冰水中，冷却后析出固体，抽滤，用冷水洗涤。粗产物用水重结晶，产量 9~10g。

2. 方法二 在 500ml 烧杯中，溶解 5ml 浓盐酸于 120ml 水中，在搅拌下加入 5.6ml 苯胺，待苯胺溶解后，再加入少量活性炭（约 1g），将溶液煮沸 5 分钟，趁热滤去活性炭及其他不溶性杂质。将滤液转移到 500ml 锥形瓶中，冷却至 50℃，加入 7.3ml 乙酸酐，振摇使其溶解后，立即加入事先配制好的 9g 结晶醋酸钠溶于 20ml 水的溶液，充分振摇混合。然后将混合物置于冰浴中冷却，使其析出结晶。抽滤，用少量冷水洗涤，干燥后称重，产量 5~6g。用此法制备的乙酰苯胺已足够纯净，可直接用于所需的合成试验。如需进一步提纯，可用水进行重结晶。

乙酰苯胺纯品为无色有闪光的小叶片状晶体，熔点为 114.3℃。本实验需 2~3h。

预 习 指 导

预习要求

1. 学习由苯胺酰化制备乙酰苯胺的原理及方法。
2. 复习分馏的原理及分馏装置的安装和操作。
3. 复习苯胺、乙酸、乙酸酐、乙酰苯胺的理化性质。
4. 复习抽滤及重结晶的基本操作。

注意事项

1. 苯胺久置易氧化变色，会影响乙酰苯胺的质量，最好用新蒸苯胺。

2. 反应中可加入少许锌粉防止苯胺被氧化，一般加入 0.1g 即可。加入过多，后处理时会生成不溶于水的氢氧化锌，增加产物处理难度。

3. 反应物冷却后，固体产物会立即析出，沾在瓶壁不易处理。故须趁热在搅动下倒入冷水中，以除去过量的醋酸和未反应的苯胺。

4. 活性炭不能加入沸腾的溶液中，否则会引起爆沸。

思考题

1. 用冰醋酸酰化实验中，反应时为什么要控制分馏柱上端的温度在 100~110℃ 之间？温度过高有什么不好？

2. 用冰醋酸做酰化试剂制备乙酰苯胺，一般可采用什么方法来提高产品的产率？

3. 在乙酸酐酰化实验中，加入盐酸和醋酸钠的目的是什么？

4. 重结晶法纯化乙酰苯胺，在制备乙酰苯胺的饱和溶液时，若烧杯中有少量油珠出现，怎样处理才算合理？试解释原因。

20 Preparation of 8-Hydroxyquinoline

Experimental Purpose

1. Master the principle of Skraup reaction and its application in the synthesis of 8-hydroxyquinoline.
2. Consolidate the basic operations of reflux, steam distillation, and recrystallization.

Experimental Principle

Quinoline and quinoline derivatives have important applications in the medicine and dye stuff industry. 8-hydroquinoline is a more important compound in quinoline derivatives, which can prepare a variety of drugs, such as iodochlorohydroxyclioquinol, chiniofon, chlorquinaldol, etc. Its divalent metal salts or salts generated with inorganic acids are the mildew-proof bactericide and algicide applying to leather, textiles, plastics, paper making, and so on.

The Skraup reaction is an important method to produce quinoline and quinoline derivatives. The reaction process is the dehydration of glycerol under concentrated sulfuric acid to produce acrolein, which is then dehydrated to form a ring after addition with 2-aminophenol. As a weak oxidant, 2-nitrophenol oxidizes 8-hydroxy-1, 2-dihydroquinoline to 8-hydroxyquinoline and reduces itself to 2-aminophenol.

HO–CH₂–CH(OH)–CH₂–OH $\xrightarrow{H_2SO_4}$ CH₂=CH–CHO + H_2O

2-aminophenol + CH₂=CH–CHO ⟶ (addition product, 3-(2-hydroxyanilino)prop-1-en-1-ol) $\xrightarrow{H_2SO_4}$ 8-hydroxy-1,2-dihydroquinoline

8-hydroxy-1,2-dihydroquinoline + 2-nitrophenol ⟶ 8-hydroxyquinoline + 2-aminophenol

Instruments and Reagents

1. Instruments round-bottom flask, still head, thermometer, thermometer adapter, condenser, receiver.

2. Reagents 2-aminephenol, 2-nitrophenol, anhydrous glycerol, concentrated sulfuric acid, sodium hydroxide solution, saturated aqueous sodium carbonate.

1. Preparation of 8-hydroxyquinoline In a dry 100ml round-bottom flask, add 9.5g of anhydrous glycerol, 1.8g of 2-nitrophenol and 2.8g of 2-aminophenol. Shake the flask vigorously until a homogenous mixture is formed. Then carefully add 9ml of concentrated sulfuric acid dropwise while cooling in an ice water bath. Install the condenser and heat the mixture until the mixture begins to reflux gently. Remove the heating source. Then heat the mixture again and bring the mixture to a gentle boil and reflux for 1.5~2 hours.

2. Purification of 8-hydroxyquinoline After slightly cooling, 15ml of water is added to remove the unreacted 2-nitrophenol by steam distillation. Cool the liquid/flask to room temperature and slowly add 6ml of sodium hydroxide solution (mass ratio 1∶1), adjust to neutral (pH=7) with saturated sodium carbonate solution. Another steam distillation is performed to produce 8-hydroxyquinoline. After cooling sufficiently, the distillate is filtered and washed, then the crystals are collected to obtain the crude product.

3. Recrystallization of 8-hydroxyquinoline Recrystallize the raw product from 25ml of a mixed solvent (ethanol: water = 4∶1V/V)to get the white needle-like crystalloid 2~2.5g. The melt point of 8-hydroxyquinoline is 75~76℃. This experiment takes about 6~8 hours.

Preview Guide

1. Review the mechanism and application of Skraup reaction.
2. Identify the substances contained in the distillate of steam distillation each time.
3. Explain the method and principle of purifying 8-hydroxyquinoline in each step.
4. Check the physical constants of the starting materials and products in the experiment.

1. The chemicals and apparatus employed in this experiment must be dried adequately.

2. All materials should be added in turn. If the concentrated sulfuric acid is added first, the reaction will be out of control.

3. Cool the crystals in the cool water to get crystals as much as possible before filter. In order to reduce the losses, wash the crystals several times in small quantities.

1. Preparation of anhydrous glycerol: put the glycerol into evaporating dish in the draught cupboard. Heat to 180℃ with an appropriate heart source, and then cool to 100℃. Then, put it into the dryer with the concentrated sulfuric acid.

2. The reaction is exothermic, and slight boiling of the solution means that the reaction has already started. In order to avoid excessive reaction, the reaction temperature needs to be controlled to a slightly

boiling state.

3. The 8-hydroxyquinoline is soluble in both acid and base to form a salt, which is not easily steamed out. The quantity of precipitation is the highest when pH of solution is controlled between 7 and 8.

Questions

1. In this experiment, why should the steam distillation be conducted in acidic condition at the first time and in neutral condition at the second time?

2. Try to explain the main factors that affect the yield and quality of the product.

实验二十 8- 羟基喹啉的合成

实验目的

1. 掌握 Skraup 反应的原理及在 8–羟基喹啉合成中应用。
2. 巩固回流、水蒸气蒸馏法、重结晶等基本操作。

实验原理

视频

喹啉和喹啉衍生物在医药和染料工业有着重要的应用。8- 羟基喹啉是喹啉衍生物中较为重要的一种化合物，可以制备氯碘喹、喹碘方、氯喹那多等多种药物。其二价金属盐或与无机酸生成的盐类是皮革、纺织品、塑料、造纸、涂料等所用的防霉杀菌剂和杀藻剂。

Skraup 反应是制备喹啉及其衍生物的重要方法，其反应过程是甘油在浓硫酸作用下脱水生成丙烯醛，再与 2- 氨基苯酚加成后脱水成环；2- 硝基苯酚作为弱氧化剂将 8- 羟基 -1, 2- 二氢喹啉氧化成 8- 羟基喹啉，本身还原成 2- 氨基苯酚。

反应式：

$$HOCH_2CH(OH)CH_2OH \xrightarrow{H_2SO_4} CH_2{=}CHCHO + H_2O$$

$$\text{2-氨基苯酚} + CH_2{=}CHCHO \longrightarrow \text{加成产物} \xrightarrow{H_2SO_4} \text{8-羟基-1,2-二氢喹啉}$$

$$\text{8-羟基-1,2-二氢喹啉} + \text{2-硝基苯酚} \longrightarrow \text{8-羟基喹啉} + \text{2-氨基苯酚}$$

仪器与试剂

1. 仪器 圆底烧瓶、冷凝管、安全管、蒸馏头、温度计、温度计套管、尾接管、接收瓶。

2. 试剂 2- 氨基苯酚、2- 硝基苯酚、甘油、浓硫酸、氢氧化钠溶液、饱和碳酸钠溶液。

实验步骤

1. 8- 羟基喹啉的制备 在 100ml 圆底烧瓶中加入 9.5g 无水甘油、1.8g 2– 硝基苯酚和 2.8g 2–氨基苯酚，混合均匀后缓慢加入 9ml 浓硫酸，安装回流装置，小火加热，微沸，撤去火源，待反应缓和至微沸时，再继续小火加热，保持回流 1.5~2h。

2. 8- 羟基喹啉的纯化 稍冷后，加入 15ml 水，进行水蒸气蒸馏，除去未反应的 2- 硝基苯酚。待烧瓶内液体冷却至室温后，缓慢加入 6ml 氢氧化钠溶液（质量比 1 : 1），再用饱和碳酸钠溶液调至中性，进行第二次水蒸气蒸馏，蒸出 8- 羟基喹啉。馏出液充分冷却后，抽滤、洗涤，收集结晶，即得粗产品。

3. 8- 羟基喹啉的精制 粗产品用 25ml 混合溶剂（乙醇 ： 水 =4 : 1 体积比）进行重结晶，得 8- 羟基喹啉纯品，约 2~2.5g。8- 羟基喹啉的熔点为 75~76℃。本实验需 6~8 小时。

预习指导

预习要求

1. 复习 Skraup 反应机理及应用。
2. 明确每次水蒸气蒸馏馏出液中所含的物质。
3. 说明纯化 8- 羟基喹啉每一步的方法及原理。
4. 查阅实验原料及产物的物理常数。

注意事项

1. 制备时所用仪器必须预先干燥。
2. 试剂要严格按所述的次序加入。如果浓硫酸先加入，则反应往往很激烈，不易控制。
3. 结晶用冷水充分冷却后再进行抽滤，保证得到尽量多的晶体；用水洗涤晶体时要少量多次，减少产物的损失。

实验说明

1. 无水甘油的制备，即将甘油在通风橱内置于蒸发皿中加热至 180℃，冷至 100℃左右，放入盛有硫酸的干燥器中备用。
2. 此合成反应为放热反应，溶液微沸即表示反应已开始，为了避免反应过于激烈，需要控制反应温度使之处于微沸状态。
3. 8- 羟基喹啉既溶于酸又溶于碱，成盐后不被水蒸气蒸馏蒸出，溶液 pH 控制在 7~8 之间，析出沉淀最多。

思考题

1. 为什么第一次水蒸气蒸馏要在酸性条件下进行，第二次却要在中性条件下进行？
2. 试说明本实验影响产率和产品质量的主要因素？

21 Preparation of Methyl Orange

Experimental Purpose

1. Learn the preparation method of diazonium salt (重氮盐).

2. Master diazonium coupling reaction (重氮化偶合反应) and the preparation method of methyl orange (甲基橙).

3. Consolidate recrystallization operation.

Experimental Principle

Diazonium salt are commonly used to prepare aromatic halogenated compounds, such as phenols, aromatic nitriles (芳香腈) and azo dyes (偶氮染料), including an indicator (指示剂) methyl orange.

Methyl orange forms beautiful orange crystals and is used as an acid-base indicator (酸碱指示剂) because of its clear and distinct color variance at different pH values. It shows red color in acidic solutions of pH less than 3.1and yellow color at pH greater than 4.4. In between, at some point there will be equal amounts of the red and yellow forms and so it presents orange.

Methyl orange is obtained by coupling the diazonium salt of *p*-aminobenzenesulfonic acid (对氨基苯磺酸的重氮盐) with acetate of *N, N*-dimethylaniline (*N, N*-二甲基苯胺的醋酸盐) in a weak acid medium.

$$C_6H_5NH_2 + H_2SO_4 \xrightarrow{170\sim180^{\circ}C} p\text{-}H_2N\text{-}C_6H_4\text{-}SO_3H + H_2O$$

$$p\text{-}H_2N\text{-}C_6H_4\text{-}SO_3H + NaOH \longrightarrow p\text{-}H_2N\text{-}C_6H_4\text{-}SO_3Na + H_2O$$

$$p\text{-}H_2N\text{-}C_6H_4\text{-}SO_3Na + NaNO_2 + 2HCl \longrightarrow p\text{-}N{\equiv}N^{+}\text{-}C_6H_4\text{-}SO_3^{-} + 2NaCl + 2H_2O$$

$$^{-}O_3S\text{-}C_6H_4\text{-}N{\equiv}N^{+} + C_6H_5\text{-}N(CH_3)_2 \xrightarrow{HOAc} [HO_3S\text{-}C_6H_4\text{-}N{=}N\text{-}C_6H_4\text{-}NH(CH_3)_2]^{+}\ OAc^{-} \longrightarrow$$

$$[HO_3S\text{-}C_6H_4\text{-}NH\text{-}N{=}C_6H_4{=}\overset{+}{N}(CH_3)_2]OAc^{-} \xrightarrow{NaOH} \underset{\text{yellow}}{NaO_3S\text{-}C_6H_4\text{-}N{=}N\text{-}C_6H_4\text{-}N(CH_3)_2}$$

Instruments and Reagents

1. Instruments three-neck flask, oil bath pot, water bath pot, drip funnel, suction filter bottle, Büchner funnel.

2. Reagents Aniline, concentrated sulfuric acid, *p*-aminobenzenesulfonic acid , sodium nitrite (亚硝酸钠), concentrated hydrochloric acid, *N, N*-dimethylaniline (*N, N*-二甲基苯胺), glacial acetic acid, 5% NaOH , ethanol, diethyl ether, starch-potassium iodide test paper (淀粉-碘化钾试纸).

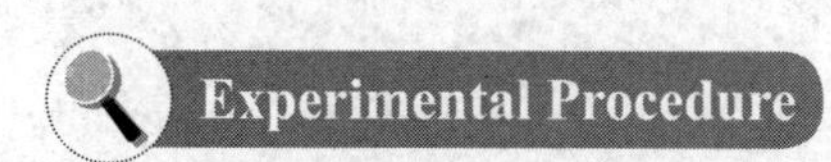

Experimental Procedure

1. Preparation of *p*-aminobenzenesulfonic acid Add 1.2g of aniline to a 50ml three-neck flask. Gradually add 2.2ml of concentrated sulfuric acid into the flask while constantly shaking and cooling with cold water. Set up the instrument with condenser tube and thermometer (the mercury bulb should be submerged in the mixture) and plug the last bottle mouth. Heat the mixture with an oil bath and maintain at a temperature of 170~180°C for 1.5 hours. Stop heating and wait until the mixture slightly cools down, then transfer the mixture into 12ml of ice water. Gray crystal of *p*-aminobenzenesulfonic acid will be precipitated under vigorous stirring. Filter the reactant with a Buchner funnel and wash it with a small amount of water to obtain a crude *p*-aminobenzenesulfonic acid product. Recrystallize it with boiling water, decolorize it with activated carbon, collect the product with suction filtration, and dry it to obtain grayish white needle-like crystals. Weigh the product and calculate the yield.

2. Preparation of methyl orange

(1) Diazotization of *p*-aminobenzenesulfonic acid Add 1.0g of *p*-aminobenzenesulfonic acid crystal and 5ml of 5% sodium hydroxide solution to a 50ml beaker. Warm the mixture to dissolve and cool it down to 0°C or less with an ice-salt bath. Prepare 0.4g of sodium nitrite in 3ml of water in a test tube and add this solution to the beaker. Prepare 1.5ml of hydrochloric acid in 5ml of water. Keep the temperature at 0~5°C and drop the dilute hydrochloric acid into the beaker while constantly stirring. When dropping is about to be finished, test the solution with starch-potassium iodide test paper until the strip turns blue. If not, add a little sodium nitrite solution until the color changes. Leave the mixture at this temperature for 15min to ensure a thorough reaction.

(2) Coupling reaction Mix 0.7ml (5.1mmol) of *N, N*-dimethylaniline and 0.5ml of glacial acetic acid in a test tube carefully. Add this solution to the cooled diazonium salt solution with constant stirring slowly. Keep stirring for another 10min. Then add about 15.5ml of 5% sodium hydroxide solution slowly until the reactant turns alkaline and has a color of orange. The crude methyl orange may precipitate as granular crystal. Heat the reactant in a boiling water bath for 5min and then shift it out. When the solution cools down to room temperature, place it in an ice bath and wait until all methyl orange precipitates. Collect the crystal with suction filtration and wash it with a small amount of water, ethanol and diethyl ether successively. Press to dry it, then weigh and calculate the yield.

Dissolve a small amount of methyl orange in water, add a few drops of dilute hydrochloric acid, and then neutralize with dilute alkaline solution. Observe the color change of methyl orange in acid and alkali. Observe the color change throughout the whole process.

The experiment takes about 8 hours.

Preview Guide

Preview Requirements

1. Review the synthesis of diazonium salt and the principle of diazonium salt coupling reaction.

2. Refer to the information and physicochemical parameters of the compounds involved in this experiment, mainly including aniline, *p*-aminobenzenesulfonic acid and N, N-dimethylaniline.

3. Review the key points of recrystallization operation.

Notes

1. The reaction temperature of aniline with concentrated sulfuric acid should be maintained at 170~180°C, if the reaction temperature is too high, it is easy to form a black viscous substance.

2. The products formed by the reaction of aniline with concentrated sulfuric acid should be poured into ice water when cooling, if the temperature falls below 50°C, the reactant will be sticky or even solidified, making it hard to pour out.

Experimental Explanation

1. *p*-aminobenzenesulfonic acid is used as an important intermediate in the preparation of azo dyes. It is an amphoteric compound with a strong sulfonic acid group and a weak amino group. The sulfanilic acid obtained from this experiment is actually in its zwitterion form.

$$H_3N^+-C_6H_4-SO_3^-$$

2. The solubility of *p*-aminobenzenesulfonic acid in water is listed below, indicating that *p*-aminoben zenesulfonic acid has a considerable solubility in water at room temperature. If the mother liquor is recovered and concentrated, an additional yield of product will be gained. After recrystallization, the product contains one molecule of crystal water and loses water at 100°C.

Temperature (°C)	Solubility (g/100ml)
100	6.67
20	1.08

3. An orange precipitate will form at this point. It is the neutral zwitterion of diazonium salt, which has low solubility at relatively low temperature.

$$^-O_3S-C_6H_4-N^+\equiv N$$

Questions

1. Why does the diazotization reaction need to be tested with starch-potassium iodide test paper?

2. Why does the temperature need to be maintained at 0~5℃ during the preparation of diazonium salt?

3. Explain the color-changing mechanism of methyl orange in acid-base medium. Represent it in a reaction equation.

实验二十一 甲基橙的合成

实验目的

1. 学习重氮盐的制备技术。
2. 掌握重氮盐的偶联反应及甲基橙的制备方法。
3. 熟悉重结晶操作。

实验原理

重氮盐常用来制备芳香卤代物、酚、芳腈及偶氮染料等，如指示剂甲基橙，它是由对氨基苯磺酸重氮盐与 *N, N*- 二甲基苯胺的醋酸盐，在弱酸介质中偶合得到的。偶合首先得到的是嫩红色的取代甲基橙，称为酸性黄；在碱性中酸性黄转变为橙黄色的钠盐，即甲基橙。甲基橙常用作染料和酸碱指示剂，其变色范围是 3.1~4.4 时呈橙色，$pH<3.1$ 时变红，$pH>4.4$ 时变黄。

甲基橙合成途径可以从苯胺开始经过磺化，进一步重氮化再与 *N, N*- 二甲基苯胺发生偶联反应制得。

反应式：

$$C_6H_5NH_2 + H_2SO_4 \xrightarrow{170\sim180℃} p\text{-}H_2NC_6H_4SO_3H + H_2O$$

$$p\text{-}H_2NC_6H_4SO_3H + NaOH \longrightarrow p\text{-}H_2NC_6H_4SO_3Na + H_2O$$

$$p\text{-}H_2NC_6H_4SO_3Na + NaNO_2 + 2HCl \longrightarrow p\text{-}N{\equiv}N^+C_6H_4SO_3^- + 2NaCl + 2H_2O$$

$$^-O_3S\text{-}C_6H_4\text{-}N{\equiv}N^+ + C_6H_5N(CH_3)_2 \xrightarrow{HOAc} [HO_3S\text{-}C_6H_4\text{-}N{=}N\text{-}C_6H_4\text{-}NH(CH_3)_2]^+\ OAc^- \longrightarrow$$

$$[HO_3S\text{-}C_6H_4\text{-}NH\text{-}N{=}C_6H_4{=}\overset{+}{N}(CH_3)_2]OAc^- \xrightarrow{NaOH} NaO_3S\text{-}C_6H_4\text{-}N{=}N\text{-}C_6H_4\text{-}N(CH_3)_2$$

yellow

仪器与试剂

1. 仪器 三颈烧瓶、油浴锅、水浴锅、滴液漏斗、抽滤瓶、布氏漏斗。

2. 试剂 苯胺、浓硫酸、对氨基苯磺酸（自制）、亚硝酸钠、浓盐酸、*N, N*-二甲基苯胺、冰醋酸、5%氢氧化钠、乙醇、乙醚、淀粉-碘化钾试纸。

实验步骤

1. 对氨基苯磺酸的制备 在50ml三颈烧瓶中加入1.2g苯胺，在冷水中冷却，不断摇动下，小心加入2.2ml浓硫酸，装上冷凝管和温度计（温度计水银球在反应液中）；另一瓶口用塞子塞住。用油浴或电热套加热，维持温度170~180℃，反应1.5小时，停止加热。稍冷，将反应物倒入12ml冰水中，在激烈搅拌下，对氨基苯磺酸呈灰白色固体析出。用布氏漏斗抽滤，少量水洗后得对氨基苯磺酸粗产品。用沸水重结晶，活性炭脱色，抽滤收集产品，晾干，得灰白色粗针状结晶，称重，计算产率。

2. 甲基橙的制备

（1）对氨基苯磺酸重氮化　在50ml烧杯中，加入1.0g对氨基苯磺酸晶体及5ml 5%的氢氧化钠溶液，温热使之溶解，用冰盐浴冷却至0℃以下。在一个试管中配制0.4g亚硝酸钠于3ml水中，将此配制溶液加入到上述烧杯中。控制温度在0~5℃，不断搅拌下，将1.5ml浓盐酸与5ml水配成的溶液缓缓滴加到上述混合液中。快滴加完时，用淀粉-碘化钾试纸检验呈现蓝色为止，若试纸不显蓝色，则补加少量亚硝酸钠溶液，直至能使淀粉-碘化钾试纸显蓝色为止。将反应液在此温度放置15min，以使反应完全。

（2）偶合反应　将量取的0.7ml (5.1mmol) *N, N*–二甲基苯胺和0.5ml冰醋酸在试管中小心混合。在不断搅拌下，将此溶液慢慢加到上述冷却的重氮盐溶液中。加完后，继续搅拌10min，然后慢慢加入15.5ml 5%氢氧化钠溶液，直至反应物呈碱性变为橙色为止。粗制的甲基橙呈细粒状晶体析出。将反应物在沸水浴上加热5min，冷却至室温后，再在冷水浴中冷却，使甲基橙晶体完全析出。抽滤，收集晶体，依次用少量水、乙醇、乙醚洗涤，压干，称重，计算产率。

将少量甲基橙溶于水中，加几滴稀盐酸，然后再用稀碱中和，观察甲基橙在酸和碱中的颜色变化。

本实验约需8小时。

预习指导

预习要求

1. 复习重氮盐的合成及重氮盐偶联反应的原理。

2. 查阅本实验涉及的化合物信息及理化参数，主要包括苯胺、对氨基苯磺酸和*N, N*–二甲基苯胺。

3. 复习重结晶操作的要点。

注意事项

1. 苯胺与浓硫酸反应温度维持在170~180℃，如果反应温度过高，容易生成黑色黏稠状物质。

2. 苯胺与浓硫酸反应生成的产物要稍冷倒入冰水，因对氨基苯磺酸当温度低于 50℃，反应物可能变粘甚至凝固，不易倒出。

实验说明

1. 对氨基苯磺酸是制备偶氮染料的重要中间体，是两性化合物，分子结构中同时存在着一个强酸性的磺酸基和一个弱碱性的氨基。本实验得到的对氨基苯磺酸实际以内盐形式存在。

$$H_3N^+-C_6H_4-SO_3^-$$

2. 对氨基苯磺酸在水中的溶解度如下。

温度（℃）	100ml 水中溶解克数
100	6.67
20	1.08

说明对氨基苯磺酸在常温下，水中的溶解度也很大，用水重结晶时，若将母液回收，浓缩还可收集部分产品。水重结晶后，产品含一分子结晶水，在 100℃ 失水。

3. 对氨基苯磺酸重氮化时有橙色沉淀析出，沉淀是重氮盐在水中电离形成的中性内盐在较低温度下水溶解度降低，析出固体结晶。

$$^-O_3S-C_6H_4-N^+\equiv N$$

思考题

1. 重氮化反应为什么需要用淀粉－碘化钾试纸检验？
2. 制备重氮盐为什么要维持 0~5℃ 的低温？
3. 解释甲基橙在酸碱介质中的变色原因。用反应式表示。

22 Preparation of 2-nitro-1, 3-Benzenediol

Experimental Purpose

1. Master the principles and experimental methods of synthesizing 2-nitro-1, 3-benzenediol (2-硝基-1, 3-苯二酚) through occupying reaction of sulfonic acid group and nitration reaction.

2. Consolidate basic operations such as steam distillation, recrystallization and vacuum filtration.

Experimental Principle

$$\text{(OH, OH)} + 2H_2SO_4 \xrightarrow{<65^\circ C} \text{(OH, OH, }HO_3S\text{, }HO_3S\text{)} + 2H_2O$$

$$\text{(}HO_3S\text{, OH, }HO_3S\text{, OH)} + HNO_3 \xrightarrow[<30^\circ C]{H_2SO_4} \text{(}HO_3S\text{, OH, }NO_2\text{, }HO_3S\text{, OH)}$$

$$\text{(}HO_3S\text{, OH, }NO_2\text{, }HO_3S\text{, OH)} + 2H_2O \xrightarrow{100^\circ C} \text{(OH, }O_2N\text{, HO)} + 2H_2SO_4$$

Phenolic hydroxyl group is not only a strong *o*-para directing group, but also a strong activating group. If resorcinol (间苯二酚) is nitrated directly, the reaction is too intense to control. In addition, due to the spatial effect, nitro group will preferentially react at the 4' and 6' positions of the benzene ring, and it is very difficult to react at the 2' position.

In this experiment, firstly sulfonic groups can be introduced at 4' and 6' positions to reduce the activity of the aromatic ring (芳香环) and occupy the active position by utilizing the strong electrophilic effect of sulfonic acid group and the reversibility of sulfonation (磺化). Secondly, according to the directing rule, the nitro-group only reacts at the second position in the nitrification (硝化) process. Finally, steam distillation is used to hydrolyze the sulfonic acid groups and steam the product with vapor. In this reaction, sulfonic acid group plays a role of occupation, localization and passivation.

The steam distillation is one of the commonly used methods to separate and purify organic compounds. It is especially suitable for the situation that products are viscous or resinous systems and difficult to purify by common distillation, extraction, crystallization and other methods.

Instruments and Reagents

1. Instruments three-necked flask, fractionating column, steam distillation unit, conical flask, beaker, Brinell funnel, suction flask, mortar, thermometer, electric furnace.

2. Reagents resorcinol, concentrated sulfuric acid, concentrated nitric acid, urea (尿素), ethanol.

Experimental Procedure

1. Synthesis of crude product Add 2.8g of resorcinol powder into a 100ml dry three-necked flask. Then slowly add 13ml of concentrated sulfuric acid and fully stir the mixture. White sulfonated product generates immediately. Heat the mixture with a hot water bath at 60~65℃ for 15min. Use ice-water bath to cool it to 25℃ or below. White sulfonated product is obtained. Add mixed acid (concentrated sulfuric acid and concentrated nitric acid, 2.8ml respectively)with a dropper. Control the reaction temperature between 25~30℃. Bright yellow paste can be obtained. Continue stirring for 15min. Carefully add 2ml of water for dilution. Control the reaction temperature at 50℃ or below. Add 0.1g of urea and then assemble a steam distillation unit for distillation (Figure 3-8). Start to distillate and orange-red solid is precipitated immediately. When there is no oil, empty the condensate and let the hot steam wash down the product accumulated in the condensing pipe. Then turn off the electric furnace to stop distillation. Cool the distillate with ice water and filter the solid to give the crude product. Then the product is dried, weighed and calculated yield.

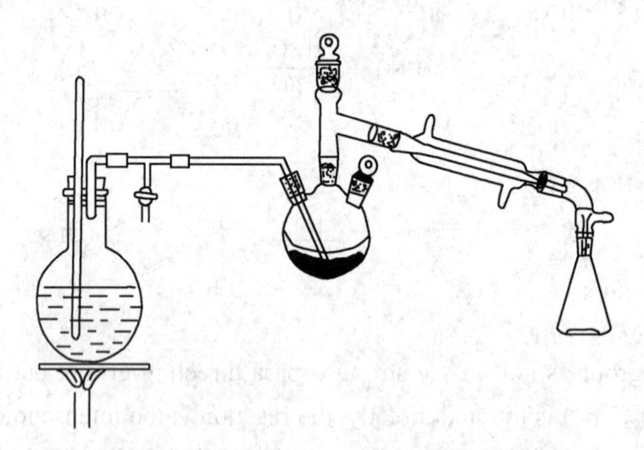

Figure 3-8 Steam Distillation Apparatus

图 3-8 水蒸气蒸馏装置图

2. Purification The crude product is recrystallized with 50% ethanol aqueous solution to obtain a refined product. Then the product is dried, weighed and the yield is calculated.

3. Determine the melting point of 2-nitro-1, 3-benzenediol, melting point 84~85℃.

Preview Guide

Preview Requirements

1. Review and consolidate the directing rule of aromatic rings and the application of protection of active sites on aromatic rings.
2. Learn the principle and operation method of sulfonation and nitration.
3. Review the principle and operation method of steam distillation experiment.

Notes

1. In this study, we must first finish sulfonation, then nitrification. Otherwise it will react violently, and even cause accidents.
2. Resorcinol must be fully pestled, or only the surface of the substrates can be sulfonated, causing incomplete sulfonation.
3. The sulfonation of phenol can be carried out at room temperature. If the reaction is too slow, or the color doesn't change white in 10 minutes, the reaction can be accelerated by heating in a water bath at 60℃.

Experimental Explanation

1. The nitration reaction is relatively fast. Therefore, before nitration, it is best to cool the sulfonated mixture and the acid mixture to 25℃ or below. Nitrification must be carried out under cooling conditions, and mixed acid should be added dropwise while stirring. Otherwise, the reaction is easily oxidized and becomes gray or black.
2. When conducting steam distillation, the condensate should be controlled very small. The reagent should be added dropwise, otherwise the product will condense at the upper end of the wall of condenser to cause blockage.

Questions

1. Why don't we do the nitration directly, but do the sulfonation first?
2. When are organic compounds purified or separated by steam distillation?

实验二十二 2-硝基-1, 3-苯二酚的制备

实验目的

1. 掌握通过磺酸基占位反应和硝化反应来合成 2-硝基-1, 3-苯二酚的原理和实验方法。
2. 巩固水蒸气蒸馏、重结晶等基本操作。

实验原理

酚羟基是较强的邻对位定位基，也是较强的致活基团。如果让间苯二酚直接硝化，由于反应太剧烈，不易控制；另外，由于空间效应，硝基会优先进入4、6位，很难进入2位。本实验利用磺酸基的强吸电子性和磺化反应的可逆性，先磺化，在4、6位引入磺酸基，即降低了芳环的活性，又占据了活性位置。再硝化时，受定位规律的支配，硝基只有进入2位，最后进行水蒸气蒸馏，即把磺酸基水解掉，又同时把产物随水蒸气一起蒸出来。本反应磺酸基起到了占位、定位和钝化的作用。

$$\text{1,3-(OH)}_2C_6H_4 + 2H_2SO_4 \xrightarrow{<65℃} \text{4,6-(HO}_3\text{S)}_2\text{-1,3-(OH)}_2C_6H_2 + 2H_2O$$

$$\text{4,6-(HO}_3\text{S)}_2\text{-1,3-(OH)}_2C_6H_2 + HNO_3 \xrightarrow[<30℃]{H_2SO_4} \text{4,6-(HO}_3\text{S)}_2\text{-2-NO}_2\text{-1,3-(OH)}_2C_6H + H_2O$$

$$\text{4,6-(HO}_3\text{S)}_2\text{-2-NO}_2\text{-1,3-(OH)}_2C_6H + 2H_2O \xrightarrow{100℃} \text{2-NO}_2\text{-1,3-(OH)}_2C_6H_3 + 2H_2SO_4$$

水蒸气蒸馏是分离和纯化有机物的常用方法之一，尤其适用于反应产物是黏稠状或树脂状体系，用一般的蒸馏、萃取、结晶等方法不易纯化的情况。

仪器设备

1. 仪器 三颈烧瓶、分馏柱、水蒸气蒸馏装置、锥形瓶、烧杯、布氏漏斗、抽滤瓶、研钵、温度计、电炉。

2. 试剂 间苯二酚、浓硫酸、浓硝酸、尿素、乙醇。

实验步骤

1. 粗产物的合成在 100ml 干燥的三颈烧瓶中，加入 2.8g 研成粉末状的间苯二酚，缓慢加入 13ml 浓硫酸，同时充分搅拌，立即生成白色的磺化物，然后在 60~65℃ 热水浴中加热 15min，然后在冰水浴中冷至 25℃ 以下，得白色糊状磺化物料。用滴管加入混酸（浓硫酸和浓硝酸各 2.8ml），控制反应温度为 25~30℃，继续搅拌 15min 后，得亮黄色糊状物。继续搅拌 15min 后，小心加入 2ml 冷水稀释，并冷却至 50℃ 以下，加入 0.1g 尿素，组装水蒸气蒸馏装置（图 3-8），进行蒸馏，馏出液中立即有橘红色固体析出。当无油状物出现，放空冷凝水，让热的蒸汽冲下积在冷凝管中的产物，即可关闭电炉停止蒸馏。冰水浴中冷却馏出液和固体，减压抽滤得粗产物，称重，计算产率。

2. 纯化粗产物用 50% 乙醇水溶液重结晶，得精制品，干燥，称重，计算产率。

3. 测定 2- 硝基 -1, 3- 苯二酚的熔点。

预 习 指 导

预习要求

1. 复习芳环定位规律和活性位置保护的应用。
2. 学习磺化、硝化反应原理及操作方法。
3. 复习水蒸气蒸馏实验原理及操作方法。

注意事项

1. 本实验一定注意先磺化，后硝化，否则，会剧烈反应，甚至发生事故。

2. 间苯二酚很硬，要充分研碎，否则，磺化只能在颗粒表面进行，磺化不完全。

3. 酚的磺化在室温就可进行，如果反应太慢，10min 不变白，可用 60℃ 的水温热，加速反应。

实验说明

1. 硝化反应比较快，因此硝化前，磺化混合物要先在冰水浴中冷却，混酸也要冷却，最好在 25℃ 以下；硝化时，也要冷却下，边搅拌，边慢慢滴加混酸，否则，反应物易被氧化而变成灰色或黑色。

2. 水蒸气蒸馏时，冷凝水要控制得很小，一滴一滴地加试剂，否则产物凝结于冷凝管壁的上端，会造成堵塞。

思考题

1. 本实验为什么不能直接硝化，而要先磺化？
2. 本实验为什么采用水蒸气蒸馏分离有机化合物？

23 Benzoin Condensation Reaction

Experimental Purpose

1. Understand the principle and method of preparing 2-hydroxy-1, 2-diphenylethan-1-one by benzoin condensation reaction.
2. Be skilled in the recrystallization operation.

Experimental Principle

Benzaldehyde (苯甲醛) is catalyzed by sodium cyanide (or potassium cyanide) and heated refluxing in ethanol. Condensation of two molecules of benzaldehyde can produce 2-hydroxy-1, 2-diphenylethan-1-one (also known as benzoin), it is benzoin condensation (安息香缩合) in organic chemistry. This mechanism is similar to the mechanism of aldol condensation reaction, which is caused by nucleophilic addition (亲核加成) of carbonyl group by carbon anion. The reaction formula is as follows.

$$2\,C_6H_5-CHO \xrightarrow[60\sim70°C]{VB_1} C_6H_5-\overset{O}{\overset{\|}{C}}-\overset{OH}{\underset{H}{C}}-C_6H_5$$

Because cyanide is highly toxic, Vitamin B_1 (Thiamine) hydrochloride instead of cyanide was used to catalyze the condensation of benzoin in this study. The reaction conditions are mild, non-toxic and high-yield.

Instruments and Reagents

1. Instruments 100ml round-bottom flask, spherical condenser, Büchner funnel, Büchner flask, conical bottle, thermostat water bath.

2. Reagents benzaldehyde, thiamine, freshly distilled benzaldehyde, 95% ethanol, 80% ethanol, 5% NaOH, activated carbon.

Experimental Procedure

Fill a 100ml round-bottom flask with 1.8g of vitamin B_1 (thiamine), 6ml of distilled water, and 15ml of 95% ethanol. Plug the bottle and chill in ice water. 5ml 5% NaOH solution is added to the test tube and cooled in ice water. The well-cooled solution of NaOH is added to the above round-bottom flask and then 10ml (0.09mol) benzaldehyde newly distilled is immediately added, then shaking thoroughly to mix the reactants evenly. The reflux condenser is installed and zeolite is added, and heated in a water bath with the

temperature controlled at 60~75℃. The reaction mixture was orange or orange-red solution. After 80~90 minutes of reaction, the water bath is removed and the reaction mixture is cooled to room temperature. Light yellow crystals are precipitated. If an oil layer appears in the reaction mixture, it should be reheated to occur a homogeneous phase, and then cooled slowly to crystallize. If necessary, glass rod can be used to rub the inner wall of round-bottom flask to promote crystallization.

After the crystallization is complete, the coarse product is filtered through a Büchner funnel, and the crystal is washed twice with 50ml cold water. The crude product can be recrystallized with 80% ethanol (if the crude product is yellow, add a small amount of activated carbon for decolorization). The output is about 4~5g.

Benzoin is a white acicular crystal with melting point of 134~136℃.

This experiment takes 7~8 hours.

Preview Guide

Preview Requirements

1. Review the reaction mechanism of aldol condensation.
2. Learn the mechanism of benzoin condensation reaction.
3. Refer to the main reagents and main physical constants of raw materials in this experiment.

Notes

1. Vitamin B_1 is stable under acidic conditions, but it is easy to absorb water, as well as it is easy to be oxidized by air and lose its function in aqueous solution and can be accelerated oxidation if exposed to light or metal ions such as Cu, Fe and Mn. In NaOH aqueous solution, the pyrimidine (嘧啶) ring of Vitamin B_1 is easy to break, so the solution of Vitamin B_1 and NaOH must be cooled with ice water before the reaction, which is the key to the success of this reaction.

2. At the beginning of the reaction, the solution does not have to boil. And later, the temperature can be appropriately raised to the slight boiling state (80~90℃).

Questions

1. It is necessary to cool in an ice bath before adding benzaldehyde in this experiment. why?
2. What is the difference about the reaction mechanism between this reaction and aldol condensation reaction?

实验二十三 安息香缩合反应

实验目的

1. 理解安息香缩合反应制备二苯乙醇酮的原理及方法。
2. 熟练重结晶操作方法。

实验原理

苯甲醛在氰化钠（钾）催化下，于乙醇中加热回流，可发生两分子苯甲醛间的缩合反应，生成二苯乙醇酮（也称安息香）。有机化学中将芳香醛进行的这一类反应都称为安息香缩合反应。该反应机理类似于羟醛缩合反应机理，是由碳负离子对羰基的亲核加成而引起的，其反应式为：

$$2\,C_6H_5-CHO \xrightarrow[60\sim70^{\circ}C]{VB_1} C_6H_5-\overset{O}{\overset{\|}{C}}-\underset{H}{\overset{OH}{C}}-C_6H_5$$

由于氰化物有剧毒，本实验改用维生素 B_1 盐酸盐代替氰化物催化安息香缩合，反应条件温和，无毒，产率较高。

仪器和试剂

1. 仪器 100ml 圆底烧瓶、球形冷凝管、布氏漏斗、抽滤瓶、锥形瓶、水浴锅。

2. 试剂 苯甲醛、维生素 B_1、95% 乙醇、80% 乙醇、5%NaOH 溶液、活性炭。

实验步骤

在 100ml 的圆底烧瓶中加入 1.8g 维生素 B_1（硫胺素）、6ml 蒸馏水和 15ml 95% 乙醇，用塞子塞住瓶口，置于冰水中冷却。取 5ml 5% NaOH 溶液加入到试管中，也置于冰水中冷却。将充分冷却的 NaOH 溶液加入上述圆底烧瓶中，并立即加入 10ml (0.09mol) 新蒸馏过的苯甲醛，充分摇动使反应物混合均匀。装上回流冷凝管，加入沸石，置于水浴锅中加热，水浴温度控制在 60~75℃。反应混合物呈橘黄色或橘红色溶液，反应 80~90min 后，撤去水浴，反应混合物冷至室温，析出浅黄色晶体，再将圆底烧瓶置冷水浴中冷却，使结晶完全。如果反应混合物中出现油层，应重新加热使其变成均相，再慢慢冷却结晶。必要时可用玻璃棒摩擦圆底烧瓶内壁，促使结晶。

结晶完全后，用布氏漏斗过滤收集粗产物，用 50ml 冷水分两次洗涤晶体。粗产物可用 80% 乙醇进行重结晶（如粗产物呈黄色，可加少量活性炭脱色）。产量 4~5g。

纯安息香为白色针状晶体，熔点 134~136℃。本实验需要 7~8h。

预 习 指 导

预习要求

1. 复习醛醇缩合的反应机理。
2. 学习安息香缩合反应机理。
3. 查阅本实验中原料的主要试剂和主要物理常数。

注意事项

1. 维生素 B_1 在酸性条件下稳定，但易吸水；在水溶液中维生素 B_1 易被空气氧化而失效，遇光或 Cu、Fe、Mn 等金属离子均可加速氧化；在 NaOH 溶液中，维生素 B_1 的嘧啶环容易开环分解，因此维生素 B_1 溶液、NaOH 溶液在反应前必须用冰水充分冷却，这是本实验成败的关键因素。

2. 反应过程中，溶液在开始时可不必沸腾，反应后期可适当提高温度至微沸状态（80~90℃）。

思考题

1. 本实验中，在加入苯甲醛之前为什么需在冰水浴中冷却？
2. 本实验的反应原理与典型的醇醛缩合反应有何不同？

Part Ⅳ
第四部分

Extraction and Purification of Natural Products and Synthesis of Simple Drugs
天然有机物提取纯化及简单药物合成

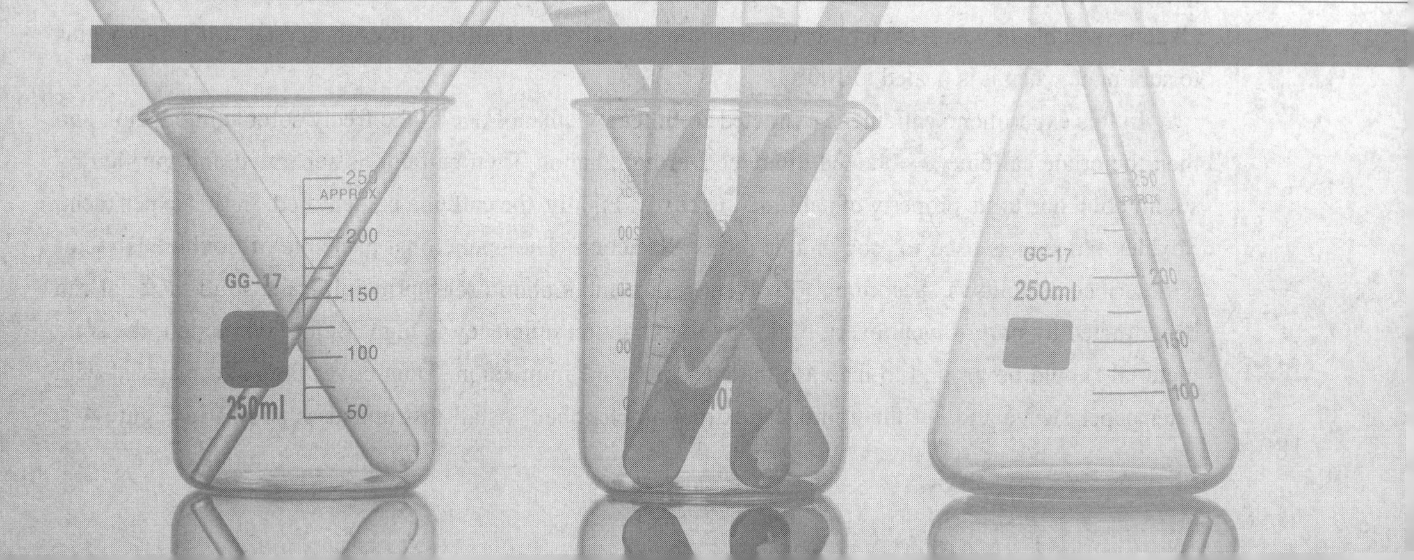

PPT

24 Extraction of Caffeine from Tea

Experimental Purpose

1. Master the operation of solid-liquid extraction with Soxhlet extractor (索氏提取器).

2. Master the principles of sublimation (升华) and use sublimation to purify solid organic compounds.

3. Know the identification method of caffeine (咖啡因).

4. Develop the students' way of thinking in extraction, separation and identification of organic compounds from natural products.

Experimental Principle

Caffeine is a derivative of heterocyclic compound purine. Its chemical name is 1, 3, 7-trimethy1-2, 6-dioxopurine. Its structural formula is as shown below.

Caffeine exists in tea, cocoa, coffee beans and so on. It is a useful active ingredient in tea and coffee for human beings. Caffeine has the effects of stimulating the heart, stimulating cerebral nerves and diuretic, so it can be used as a central nervous stimulant. Caffeine is an alkaloid, and tea contains a variety of alkaloids. Caffeine, as the main alkaloid in tea, accounts for about 1%~5% of tea quality. In addition, tea also contains 11%~12% tannic acid, 0.6% pigment, cellulose, protein and so on. Caffeine is a white needle-shaped crystal, usually containing one molecule of crystal water ($C_8H_{10}O_2N_4 \cdot H_2O$). It is weak alkaline, soluble in water, ethanol, benzene and other solvents. Caffeine loses its crystal water and begins to sublimate when it is heated to 100℃.

In this experiment, caffeine is extracted from tea by ethanol due to the freely soluble in ethanol, and then the crude caffeine is obtained after recovering ethanol. Then caffeine is separated and purified by sublimation due to its property of sublimating easily. Finally, the caffeine is identified. In this experiment, Soxhlet extractor is used for continuous reflux extraction. The operational principle of Soxhlet extractor is described as follows. According to solvent reflux and siphon (虹吸)principles, the solid material can be extracted by pure solvent every time, so the extraction efficiency is high. Before extraction, the solid material should be ground to increase the area of liquid immersion. Then cover the solid material with filter paper sleeve and put them into the extraction chamber. Install instrument as shown in Figure 4-1.

When the solvent is heated to boiling, the vapor rises through the air duct and then condensed into liquid to drip into the extractor. When the liquid level exceeds the highest point of the siphon, siphoning occurs and the solution goes back into the flask, so that it can extract some soluble substances in the solvent. The soluble solids in the flask are enriched by utilizing solvent reflux and siphon principles.

Instruments and Reagents

1. Instruments Soxhlet extractor, spherical condenser tube, round-bottom flask, straight condenser tube, connecting tube, conical flask, glass funnel, evaporating dish, mortar, test tube, waterbath, sand bath, micro melting point instruments.

2. Reagents tea leaves, 95% ethanol, CaO powder, river sand and distilled water.

Experimental Procedure

1. Extraction Weigh 10g of tea leaves into a suitable filter paper sleeve, then put it into a Soxhlet extractor. Add 100ml of 95%ethanol and boiling chips into a flat bottom flask. Prepare a reflux instruments with a water bath. Reflux and extract for 1~2 hours continuously until the color of extracted liquid is very light. Stop heating instantly once the liquid just siphons back to the flask from the Soxhlet extractor (Figure 4-1).

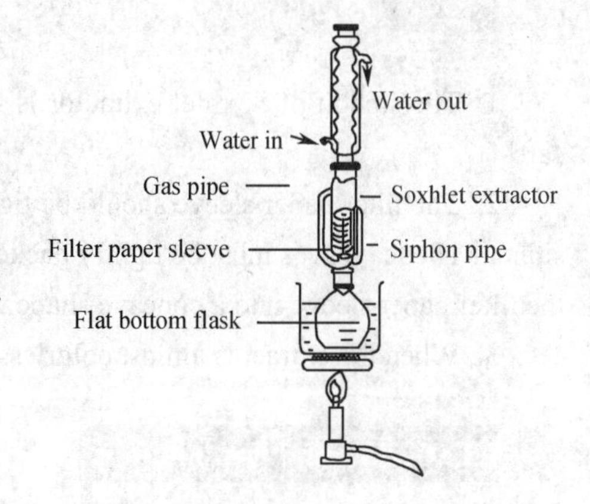

Figure 4-1 Installation Drawing of Extractor for Caffeine

图 4-1 咖啡因提取装置图

2. Distillation and concentration Use a distillation unit to recover most of the ethanol (pour it into a recovery bottle) and then pour the residual liquid (about 10ml)into an evaporating dish.

3. Alkali treatment Add 3~4g of CaO powder. Place the evaporating dish on the steam bath, and evaporate the solvent under the constant stirring with a glass rod until the solid turns into granular powder. After cooling, wipe off the powder stuck on the edge to avoid contaminate the product when sublimated.

4. Sublimation Take a glass funnel with the same size as the evaporating dish, and tuck a loose mass of cotton in the neck of the funnel. Cover the funnel on the evaporating dish with filter paper (the filter paper should be larger than the evaporating dish, and the part covered by the glass funnel should be perforated)and heat the evaporating dish with hot sand bath for caffeine sublimation. Stop heating when white smoke comes out of the small hole and yellow oil appears on the funnel. Use the crucible tongs carefully to remove the entire sublimation Instruments from the sand bath and cool it for about 5 minutes. Collect the caffeine on weighing paper. If the residue is green, it can be sublimated again after mixing, so that it can be collected completely. The combined caffeine is weighed and the extraction rate can be calculated.

5. Identification

(1) Dissolve a little caffeine in dichloromethane. Use a capillary tube to test the liquid on alkaline or neutral silica gel plate. Spread out the plate with different developing solvents, and spray bismuth potassium iodide test solution on it. See the development of its spots to determine its purity.

(2) Determine the melting point of pure caffeine, with a reference value of 236°C.

Preview Guide

Preview Requirements

1. Learn the principle and operation of solid-liquid extraction with Soxhlet extractor.

2. Learn the principle of sublimation and operation method and the application scope of sublimation.

3. Compare the similarities and differences between solid-liquid extraction and liquid-liquid extraction.

Notes

1. The siphon of Soxhlet extractor is very thin and easy to break, so it should be used with special care.

2. The filter paper sleeve should be tightly attached to the wall and the height should not exceed the siphon. The tea leaves must be tightly packed to prevent leakage and blockage of siphons. Fold the top of the filter paper sleeve into a concave shape to ensure the reflow evenly moistens the extract.

3. When the extract is almost colorless, the extraction can be stopped.

Experimental Explanation

1. The ethanol in the flask cannot be distilled too dry, otherwise the residue is too sticky and difficult to transfer to cause losses.

2. The residual solution is treated with quicklime to absorb moisture and neutralize acidic substances. Natural alkaloids generally exist in the form of salt, which is easy to sublimate after neutralization.

3. The success of this experiment depends on the sublimation. Heat slowly and indirectly in the sublimation process. The sand bath temperature should not be too high (the best sand bath temperature is 300℃ or so), otherwise it is easy to make the filter paper carbonize, and some colored materials are baked out, affecting the yield and purity.

Questions

1. Explain the principle of Soxhlet extractor briefly. What are the advantages of using it in solid-liquid extraction?

2. What is the effect of adding CaO powder in caffeine extraction?

3. Why do we use steam bath instead of heating directly to evaporate the sample?

实验二十四　从茶叶中提取咖啡因

实验目的

1. 掌握用索氏提取器进行固–液萃取的操作。
2. 掌握升华的原理并用升华进行固体有机化合物提纯的操作方法。
3. 了解咖啡因的鉴定方法。
4. 培养学生从天然产物中提取、分离和鉴定有机化合物的思维方法。

实验原理

咖啡因又称咖啡碱，是杂环族化合物嘌呤的衍生物，它的化学名称是 1, 3, 7- 三甲基 -2, 6- 二氧嘌呤，其结构式如下：

咖啡因存在于茶叶、可可、咖啡豆等植物体内，是茶叶和咖啡中对人类有用处的活性成分。咖啡因具有刺激心脏、兴奋大脑神经和利尿等作用，因此可作为中枢神经兴奋药。咖啡因是一种生物碱，茶叶中所含的多种生物碱，以咖啡碱为主，占茶叶质量 1%~5%。此外，茶叶中还含有 11%~12% 的丹宁酸（又称鞣酸），0.6% 的色素、纤维素、蛋白质等。咖啡碱为白色针状结晶，通常含有一分子结晶水（$C_8H_{10}O_2N_4 \cdot H_2O$），弱碱性，易溶于水、乙醇、苯等溶剂。咖啡碱受热到 100℃ 时即失去结晶水，并开始升华。

本实验利用咖啡因易溶于乙醇的性质，采用乙醇提取茶叶中的咖啡因，回收乙醇后得到富含咖啡因的粗品，再利用咖啡因易于升华的性质进行分离和提纯，最后进行鉴定。

本实验采用索氏提取器连续回流提取。索氏提取器的工作原理是利用溶剂回流和虹吸原理，使固体物质每一次都能为纯的溶剂所萃取，所以萃取效率较高。萃取前应先将固体物质研磨细，以增加液体浸溶的面积。然后将固体物质放在滤纸套内，放置于萃取室中。如图 4-1 安装仪器。当溶剂加热沸腾后，蒸气通过导气管上升，被冷凝为液体滴入提取器中。当液面超过虹吸管最高处时，即发生虹吸现象，溶液回流入烧瓶，因此可萃取出溶于溶剂的部分物质。就这样利用溶剂回流和虹吸作用，使固体中的可溶物富集到烧瓶内。

仪器和试剂

1. 仪器　索氏提取器、球形冷凝管、平底烧瓶、直形冷凝管、接引管、锥形瓶、玻璃漏斗、

蒸发皿、研钵、试管、水浴、沙浴、显微熔点测定仪。

2. 试剂 茶叶、95% 乙醇、生石灰粉、河沙、蒸馏水。

1. 提取 称取 10g 茶叶末放入索氏提取器的滤纸套中，在平底烧瓶内加入 95% 乙醇 100ml，加热萃取，连续提取 1~2h（提取液颜色很淡时，即可停止提取）后，待冷凝液刚刚虹吸下去时，立即停止加热（图 4-1）。

2. 蒸馏浓缩 改用蒸馏装置，蒸出大部分乙醇（倒入回收瓶），再把残余液（约 10ml）倒入蒸发皿中。

3. 碱处理 拌入 3~4g 生石灰粉。将蒸发皿放在水蒸气浴上，在玻璃棒不断搅拌下蒸干溶剂，搅拌直至固体变成小颗粒粉末。冷却后，擦去沾在边上的粉末，以免在升华时污染产物。

4. 升华 取一只与蒸发皿大小一致的玻璃漏斗，漏斗颈部塞一团疏松的棉花，将其罩在铺有滤纸的蒸发皿上（滤纸的直径应大于蒸发皿，其被玻璃漏斗罩住的部分应扎满小孔），用沙浴加热使其升华。当从小孔中冒出白烟，漏斗上出现黄色油状物时停止加热。用坩埚钳小心地把整个升华装置移出沙浴，冷却 5min 左右，取下漏斗和滤纸，将咖啡因用小刀刮到称量纸上。残渣如果为绿色可经搅拌后再次进行升华，使之收集完全。合并所得的咖啡因，并称重，计算提取率。

5. 鉴定 ① 取少许咖啡因，溶于二氯甲烷中，用毛细管将此液点样于碱性或中性硅胶板上，用不同的展开剂展开，再喷碘化铋钾试液，看其斑点的展开情况，以确定其纯度。② 测定纯净咖啡因的熔点，文献值为 236℃。

预 习 指 导

预习要求

1. 学习索氏提取器进行固 - 液萃取的原理及操作方法。
2. 学习升华原理、操作方法以及升华适用范围。
3. 比较固-液萃取和液-液萃取的异同点。

注意事项

1. 索氏提取器的虹吸管很细，非常容易折断，仪器装置和使用时须特别小心。

2. 滤纸套大小以紧贴器壁，高度不超过虹吸管为宜。滤纸包茶叶末时要严实，以防漏出堵塞虹吸管。滤纸套上面折成凹形，确保回流液均匀湿润萃取物。

3. 提取液接近无色时，即可停止提取。

实验说明

1. 蒸馏时烧瓶中的乙醇不可蒸得太干，以免残渣太黏，转移困难且损失大。

2. 残液用生石灰粉处理，主要是吸水和中和，以除去酸性物质。天然生物碱一般以盐的形式存在，中和后游离出来易于升华。

3. 本实验成功与否取决于升华操作。在升华过程中始终用小火间接加热，沙浴温度不可过高

（最好沙浴温度设定在300℃左右），否则易使滤纸炭化变黑，并把一些有色物质烘出来，影响产率和纯度。

思考题

1. 简要说明索氏提取器进行固–液萃取的工作原理。用索氏提取器进行固–液萃取有何优点？

2. 提取咖啡因时加入生石灰起什么作用？

3. 蒸干水分时为什么要用水蒸气浴而不直接加热？

25 Extraction of Berberine From Rhizoma Coptidis

Experimental Purpose

1. Be familiar with the property and application of berberine (黄连素).
2. Master the principle and method of extracting berberine from rhizoma coptidis (黄连).
3. Learn and master the chemical identification and TLC identification of alkaloids.

Experimental Principle

Berberine, an isoquinoline alkaloid (异喹啉类生物碱), is the main component of rhizoma coptidis, the content of which can reach 4%~10%. In addition to rhizoma coptidis, Chinese herbs such as cortex phellodendron (黄柏), berberis (三颗针), felled cattle flower (伏牛花) and celandine (白屈菜) also contain berberine, among which rhizoma coptidis and rhizoma cypress have the highest content. Berberine has antibacterial, anti-inflammatory, anti-diarrhea and other effects, which is widely used in clinical.

Berberine has three tautomeric isomers (互变异构), but most of them exist in nature as quaternary ammonium base (季铵碱). Berberine is slightly soluble in water and ethanol, more soluble in hot water and hot ethanol, almost insoluble in ether. Berberine hydrochloride, sulfate, hydroiodate and nitrate are insoluble in cold water, but soluble in hot water. Berberine is extracted from rhizoma coptidis with appropriate hot solvent (such as ethanol, water, sulfuric acid aqueous solution, etc). The crude product can be purified by recrystallization with water.

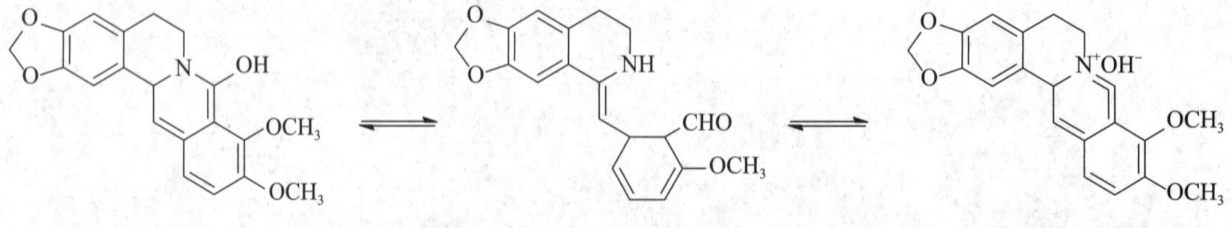

（醇式） ⇌ （醛式） ⇌ （季铵碱式）

Berberine is oxidized by oxidizing agents such as nitric acid and then converted into fuchsia oxide berberine. Part of berberine is converted into aldehyde-berberine in a stronger base condition, and the condensation reaction occurs between acetone and aldehyde-berberine to produce a yellow precipitation.

Instruments and Reagents

1. **Instruments** round-bottom flask, spherical condenser.
2. **Reagents** rhizoma coptidis, ethanol, 1% acetic acid, hydrochloric acid, sulfuric acid.

1. Extraction of berberine Weigh 15g of rhizoma coptidis, pulverize (研磨) into powder, then put it into a 250ml round-bottom flask and add 150ml ethanol, which install a reflux condenser. Heat and reflux for 0.5h, cool and soak for 0.5h, then filter out the solid residue. The filter residue is repeated the above operation twice. Combine the filtrate for three times, and the ethanol is recovered by vacuum distillation.

2. Purification of berberine Add 30ml 1% acetic acid to the concentrated solution. Remove solid impurities by hot filtration after heating and dissolving. Then add hydrochloric acid to the filtrate until the solution is cloudy (about 10ml). When cooled in an ice bath, yellow acicular berberine hydrochloride crystals are precipitated. The crude product of berberine hydrochloride can be obtained by washing twice with ice water after filtration and crystallization.

3. Put the above crude berberine into a 100ml beaker, add 30ml water, heat, stir, and boil for a few minutes, and then filter it while it is hot. Adjust the filtrate with hydrochloric acid to pH 2~3. Leave it at room temperature for several hours, then the orange-yellow crystallization is precipitated. The crystal is separated by filter and the filter residue is washed twice with a little cold water to obtain the final product.

4. Qualitative identification of berberine

(1) Chemical identification ① Take the sample about 50mg, add 5ml distilled water, slowly heat the system to dissolve, and then add 2 drops of 20% NaOH solution to produce orange color. The solution is cooled and filtered. Added a few drops of acetone to the filtrate, and then lemon yellow acetone berberine precipitation is occurred. ② Take a few samples, add 2ml of concentrated sulfuric acid, then add a few drops of concentrated nitric acid to the solution to display a fuchsia (紫红色) solution.

(2) TLC identification Prepare a thin layer chromatography plate of alumina. Take a small amount of berberine, which is dissolved in 2ml ethanol (if necessary, it can be heated briefly in a water bath). Gently draw a straight line with a pencil 2cm from one end of the TLC plate, then dot the sample with the flat capillary (毛细管) tube at the pencil line. Carefully place the plate with the well-pointed sample into the chromatography cylinder, and unfold with chloroform-methanol (9 : 1) as the developer. About 1cm from the front edge of the agent to be developed to the upper end of the TLC plate, remove the TLC plate, mark the front edge with a pencil, dry, and calculate the R_f value of berberine.

Preview Guide

Preview Requirements

1. Understand the basic principle and experimental process of extraction and purification of berberine.
2. Learn the principle of chemical identification of berberine.
3. Review the basics operation of reflux, learn recrystallization and TLC.

Notes

1. In addition to rhizoma coptidis, cortex phellodendron can also be used in this experiment.

2. In the raw materials such as rhizoma coptidis, except mainly containing berberine, there are still a certain amount of root alkaloid, bating and other components. These components content is small, so there is no need to separate.

3. The concentration of dilute sulfuric acid used for extraction is between 0.2%~0.3%. If the concentration of dilute sulfuric acid is increased, berberine will change from sulfate to berberine sulfate (B · HSO_4)-acid salt, whose solubility (1 : 100) is significantly lower than sulfate (1 : 30), which affecting the extraction effect.

4. The purpose of adding NaCl is to reduce the solubility of berberine hydrochloride in water by using its salting effect.

5. In the process of preparation the crude berberine hydrochloride, the solution after boiling should be quickly filtered while hot, so as to avoid the solution cooling and precipitation of berberine hydrochloride crystals, resulting in a lower extraction rate.

Experimental Explanation

1. Berberine can also be extracted by Soxhlet extractor.

2. It is difficult to obtain the pure berberine crystals. Heating the berberine hydrochloride in water until just dissolve and bring to a boil. Adjust pH of the solution with lime milk from 8.5 to 9.8, and then cool down and filter impurities, then continue to cool the filtrate below room temperature and the acicular-like berberine precipitation is obtained. The melting point of berberine is 145℃.

3. If the temperature is too high and the solution boils violently, pectin-like substances are also extracted from the rhizoma coptidis, which will make subsequent filtration be difficult.

4. When spotting, capillary liquid level just contact the TLC plate, do not overweight and destroy the thin layer.The liquid level must be below the dot line, and do not beyond the dot line.

Questions

1. What kind of alkaloid is berberine?

2. Why should adjust pH with lime milk? Using strong alkali, such as NaOH (or KOH), is it OK or not?

实验二十五　从黄连中提取黄连素

实验目的

1. 熟悉黄连素的性质和应用。
2. 掌握从中药黄连中提取黄连素的原理和方法。
3. 学习和掌握生物碱的化学检测及薄层色谱的鉴定方法。

实验原理

黄连素（也称小檗碱），属于异喹啉类生物碱，是中药黄连的主要成分，含量可达4%~10%。除黄连外，黄柏、三颗针、伏牛花、白屈菜等中草药中也含有黄连素，其中以黄连和黄柏中含量最高。黄连素有抗菌、消炎、止泻等功效，在临床上有广泛应用。

黄连素存在三种互变异构体，但自然界多以季铵碱形式存在。黄连素微溶于水和乙醇，较易溶于热水及热乙醇中，几乎不溶于乙醚。其盐酸盐、硫酸盐、氢碘酸盐、硝酸盐均难溶于冷水，易溶于热水。从黄连中提取黄连素，往往采用适当的热溶剂（如乙醇、水、硫酸等）提取，然后浓缩，再加酸进行酸化，得到相应的盐。粗产品可用水重结晶进行提纯。

OH　OCH_3　OCH_3　⇌　NH　CHO　OCH_3　⇌　N^+OH^-　OCH_3　OCH_3

（醇式）　（醛式）　（季铵碱式）

黄连素被硝酸等氧化剂氧化，转变为樱红色的氧化黄连素。黄连素在强碱中部分转化成醛式黄连素，在此条件下，加几滴丙酮，即可发生缩合反应，丙酮与醛式黄连素缩合产物为黄色沉淀。

仪器和试剂

1. 仪器　圆底烧瓶、球形冷凝管。

2. 试剂　黄连、乙醇、1%醋酸、浓盐酸、硫酸。

实验操作

1. 黄连素的提取

（1）黄连素的提取　称取15g黄连，粉碎成末，放入250ml圆底烧瓶中，加入150ml乙醇，装上回流冷凝管，加热回流0.5h，冷却，静置浸泡0.5h，抽滤。滤渣重复上述操作两次，合并三

次滤液，减压蒸馏回收乙醇，得棕红色糖浆状物。

（2）黄连素的纯化　① 浓缩液中加入 1% 醋酸 30ml，加热溶解后趁热抽滤除去固体不溶物，然后在滤液中滴加浓盐酸至溶液浑浊为止（约需 10ml）。冰水浴中冷却，即有黄色针状的黄连素盐酸盐晶体析出。抽滤，结晶用冰水洗涤两次，得黄连素盐酸盐粗产品。② 将上述黄连素粗品放入 100ml 烧杯中，加入 30ml 水，加热至沸腾，搅拌沸腾几分钟，趁热抽滤，滤液用盐酸调节至 pH 2~3，室温放置数小时，有较多橙黄色结晶析出，抽滤，滤渣用少量冷水洗涤两次，烘干即得成品。

2. 黄连素的定性鉴别

（1）化学检识　① 丙酮加成反应：取样品约 50mg，加 5ml 蒸馏水，缓缓加热溶解，加 20% NaOH 溶液 2 滴，溶液变为橘红色，放冷后过滤，滤液中加丙酮数滴，即有柠檬黄色的丙酮黄连素沉淀析出。② 氧化显色反应：取样品少许，加浓硫酸 2ml，溶解后加几滴浓硝酸，即呈樱红色溶液。

（2）薄层色谱检识　制备氧化铝薄层板，取少量黄连素结晶溶于 2ml 乙醇中（必要时可在水浴上加热片刻）。在离薄层板一端 2cm 处用铅笔轻轻划一直线，取管口平整的毛细管插入样品溶液中取样，并在铅笔划线处轻轻点样，将点好样品的薄层板小心放入展开槽内，以三氯甲烷 - 甲醇（9 : 1）为展开剂进行展开。待展开剂前沿距薄层板上端约 1cm 处，取出薄层板，用铅笔在前沿划一记号，晾干，计算黄连素的 R_f 值。本实验需 6~7h。

预 习 指 导

预习要求

1. 理解黄连素的提取、纯化的基本原理及实验流程。
2. 学习黄连素的化学鉴别原理。
3. 复习回流、重结晶、薄层色谱等基本操作。

注意事项

1. 实验材料除可用黄连外，也可用黄柏皮。

2. 黄连等原料中除主要含小檗碱外，尚含有一定量的小檗胺、药根碱和巴马亭等多种成分，除小檗碱、小檗胺含量多且有一定药用价值外，其余成分含量少，且无分离必要。

3. 提取用稀硫酸浓度以 0.2%~0.3% 之间为宜，若加大稀硫酸浓度，小檗碱将会从硫酸盐转变成硫酸氢小檗碱（$B \cdot HSO_4$）- 酸式盐的形式，后者的溶解度（1 : 100）明显地较硫酸盐（1 : 30）小，影响提取效果。

4. 加 NaCl 的目的是利用其盐析作用以降低盐酸小檗碱在水中的溶解度。

5. 粗制盐酸小檗碱过程中，煮沸后的溶液应趁热迅速抽滤，以免溶液冷却而析出盐酸小檗碱结晶，造成提取率降低。

实验说明

1. 提取回流装置也可用索氏提取器（也称为脂肪提取器）。

2. 得到纯净的黄连素晶体比较困难。将黄连素盐酸盐加热水至刚好溶解煮沸，用石灰乳调节

pH 8.5~9.8，冷却后滤去杂质，滤液继续冷却至室温以下，即有针状的黄连素析出，抽滤，将结晶在 50~60℃ 下干燥，熔点 145℃。

3. 如果温度过高，溶液剧烈沸腾，则黄连中的果胶等物质也被提取出来，使得后面的过滤难以进行。

4. 点样时，毛细管液面刚好接触薄层即可，切勿点样过重而破坏薄层。

5. 展开剂液面一定要在点样线下，不超过点样线。

思考题

1. 黄连素为何种生物碱？
2. 为何要用石灰乳来调节 pH，使用强碱氢氧化钠（钾）行不行？

26 Extraction of Volatile Oil

Experimental Purpose

1. Learn the extraction principle and method of traditional Chinese medicine volatile oil.
2. Master the procedure of steam distillation and how to use essential oil extractor.
3. Learn the idea and method of experimental design.

Experimental Principle

Volatile oil (挥发油), also known as essential oil (精油), is defined as oily liquid present in plants possessing volatility, distillable with steam and water-insoluble. Most volatile oils have aromatic smell. Volatile oil is one of the most important active ingredients, mainly used in medicine and health care products, cosmetics and food industries.

Volatile oils contain a variety of types ingredients and are usually comprised of dozens to hundreds of constituents, in which one or more ingredients account for larger. They are usually consisted of aliphatic, aromatic compounds, and terpenoids.

Volatile oils are normally colorless or light-yellow oily liquid. Most volatile oils are lighter than water, while some of them, such as clove oil and cassis oil, are heavier than water. Volatile oils are soluble in ether, chloroform, petroleum ether, carbon disulfide, fatty oils, and other lipophilic solvents, and practically insoluble in water. Volatile oils are usually soluble completely in anhydrous ethanol, but the solubility decreases with the declination of concentration of ethanol. When volatile oils are mixed with fatty oils or terpenes, the solubility in a certain concentration of ethanol will decrease. Therefore, the purity of volatile oil can be tested with the solubility of volatile oil in alcohol according to the *Pharmacopoeia of the People's Republic of China* (中华人民共和国药典).

Volatile oils are extracted by distillation, solvent extraction and supercritical fluid extraction. Steam distillation is the common method, as the boiling point of the mixture is below the boiling point of any component in the mixture, that is, when the water is very sufficient, volatile oils and water will be distilled together at a temperature lower than 100℃. This extraction method can not only increase the distillation speed of volatile oil, but also ensure that during operation the temperature will not cause the raw materials to easily become decomposed, which will affect the yield and quality of volatile oil.

Chromatography, especially gas chromatography, is mainly used for analysis of volatile oil.

26.1 Extraction of Paeonol from Cortex Moutan

PPT

Tree Peony bark is the root bark of the buttercup peony, which is slightly cold in nature and tasted bitterly. It has the effect of clearing heat from blood and activating blood circulation to dissipate blood stasis.

The main ingredients of Tree Peony bark are Paeonol (cont 1.9%~2.3%) and paeonoside, etc. As the latter can easily decompose, it produces paeonol during storage. In addition, the root of *Cynanchum paniculatum (Bunge) Kitagawa* (徐长卿) also contains a lot of paeonol (芍药醇).

Paeonol is an aromatic, white acicular (针状的) crystal (mp 50°C), which is effective on analgesia, sedation and anti-bacteria. Clinically, it can treat rheumatism, toothache, stomachache, skin disease, chronic bronchitis (支气管炎) and asthma (哮喘).

Paeonol of chemical name is 2-hydroxy-4-methoxy acetophenone. The formulas are as follows:

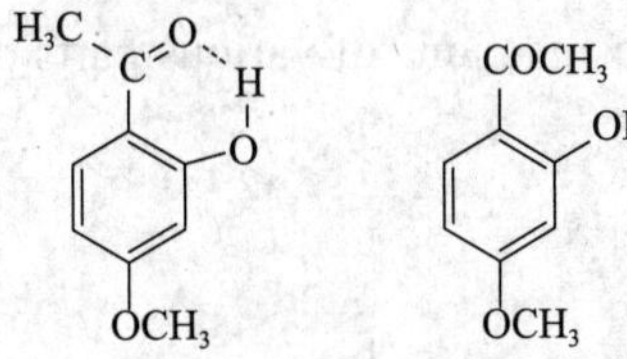

In the structure of paeonol, the intra-molecular hydrogen bond can be formed between both the carbonyl and hydroxyl groups in the ortho position. Therefore, it can be extracted by steam distillation as its volatility. Paeonol can be purified according to its insolubility in cold water and solubility in ethanol, ethyl ether, chloroform, benzene, etc.

Instruments and Reagents

1. Instruments electronic scales; steam distiller; melting point instruments; beaker; measuring cylinder; vacuum filter.

2. Reagents Cortex Moutas; 1%$FeCl_3$; NaCl; 95% EtOH; Iodine liquid; NaOH.

Experimental Procedure

1. Extraction 30g of crushed Tree Peony bark or *Cynanchum paniculatum (Bunge) Kitagawa* root, and 1g of NaCl are placed in a 500ml flask, then an appropriate amount of hot water is added to moisten the powder. In the steam distillation device, 5g of NaCl is in a 250ml beaker as the receiving container, which is cooled in an ice-water bath. Steam is pumped into the flask for distillation. Distillation will be stopped when the distillate becomes transparent without opacification. Afterwards, the distillate will be kept on cooling in an ice-water bath for complete solidification.

2. Separation and Purification After standing fully, crude product can be obtained by vacuum filtering. Dissolve the crystals with a small amount of 95% ethanol (less than 5ml), then add a large amount of distilled water (approximately 4 times the amount of ethanol solution). The solution is milky white at first, and a large number of white, acicular crystals are precipitated after standing. The pure

paeonol are obtained through filtering and natural drying.

3. Identification

(1) Iodoform (碘仿) experiment Preparation of paeonol-methanol solution. Beige precipitate will appear in iodoform reaction

(2) Ferric chloride ($FeCl_3$) test Preparation of paeonol-ethanol or methanol solution; The solution is purplish red when $FeCl_3$ is added.

(3) Melting point determination mp 50°C.

Preview Guide

1. Know the function and uses of paeonol.
2. Recognize and write down the molecular structure of paeonol.
3. Be familiar with the names and functions of various parts of steam distiller, and operations need to be paid more attention.

1. The contents of paeonol in the bark of peony vary greatly in different seasons and regions, higher in Spring and in Sichuan.
2. In the process of steam distillation, it is theoretically necessary to display a colorless state with ferric chloride, which indicates a finished steam distillation, that is, there is no positive reaction of paeonol. However, it may take longer time and be less efficient, so the appearance of clear colorless distillate means the end point.

1. The operation of steam distillation is detailed in Experiment 7.
2. If white crystals are not obtained, but the pearl-like droplet sinks in the distillate during the extraction, steps can be taken to grow the crystal: adding some paeonol crystal seed or scraping the inside wall of the beaker, that a quality of white acicular crystals is precipitated. The distillate could also be extracted three times by shaking with ether (30ml, 20ml, 15ml), and combined ether solution is dehydrate with anhydrous sodium sulfate. A small amount of ether keeps standing overnight, and the white crystals precipitate out.
3. NaCl can obviously increase the extraction rate of paeonol and shorten the extraction time. In reports, the amount of NaCl ranges from 5% to 10% of Moutan Bark. Paeonol has a certain solubility in hot water. A certain concentration of NaCl in the extract will reduce the dissolvability of Paeonol in water, meanwhile accelerate its precipitation.

1. Why is paeonol volatile based on the chemical structure?

2. Can ethanol be used as a solvent for paeonol testing during chemical detection (iodoform test) ? Why is that?

3. Can salicylic acid (水杨酸) also be extracted and separated by steam distillation? Why?

4. Based on what you've learned, please give a few of familiar examples that can be extracted or separated by steam distillation.

26.2 Extraction of Cinnamaldehyde from Cinnamon

Cinnamaldehyde (肉桂醛), i. e. trans-3-phenylacrolein, is an aldehyde organic compound, which is abundant in cinnamon. The cinnamaldehyde is trans structure in nature, and its structure is as follows.

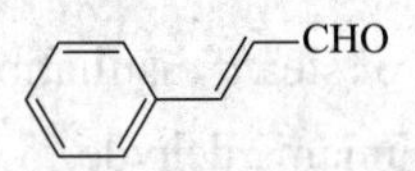

cinnamaldehyde

Cinnamaldehyde is mainly used as a perfumer in drinks and food, and also used in other products. It is colorless or light-yellow liquid with density of 1.046~1.052, melting point of –7.5℃, index of refractive 1.619~1.623, and boiling point of 253℃ at 760 torr. It is insoluble in water, and soluble in alcohol, ether, dichloromethane and other organic solvents. It also evaporates with steam.

It is unstable in strong acidic or alkaline medium, and can be oxidized to cinnamic acid in air.

The methods of extracting cinnamaldehyde include steam distillation, pressing, and solvent extraction. Steam distillation is used in this experiment. The extraction yield is about 2.0%.

1. Instruments 50ml round-bottom flask, three-way adapter, thermometer, thermometer adapter, condenser, adapter, receiving flask, separating funnel.

2. Reagents cinnamon peel, methylene chloride, sodium sulfate anhydrous, Tollens reagent, Schiff reagent, Br_2/CCl_4 solution.

1. Extraction of cinnamaldehyde After weighed, grinded in a mortar, 10g cinnamon peel together with 20ml ware are put into a 50ml round-bottom flask, then kept refluxing for 20minutes. After cooled down to room temperature, the steam distillation equipment is set up and kept heating then 8~10ml distillate is collected. The distillate is transferred to a 20ml separating funnel and extracted with dichloromethane for three times (3 × 2ml). Water layer is removed, and the dichloromethane is combined in a test tube. A small amount of anhydrous sodium sulfate is added into the test tube, and 20minutes later, the extraction solution is decanted (倒出), heated and steamed in a water bath in the fume hood and cinnamaldehyde is got.

2. Property test of cinnamaldehyde

(1) 1 drop of extract and 1 drop of Br_2/CCl_4 solution are taken into a test tube, then the color fades or not.

(2) 2 drops of the extract and 2 drops of 2, 4-dinitrophenylhydrazine (2, 4-二硝基苯肼) reagent are added into a test tube, and there is yellow precipitation or not.

(3) 1 drop of the extract and 2 drops of Tollens reagent are put in a test tube, heated in a water bath, and there is appearance of silver mirror or not.

(4) 1 drop of the extract and 2 drops of Schiff reagent are put in a test tube, and shake for 1 min, there is purple red or not. If purple red does not appear, then taken a water bath and slightly heated for 2~3min.

Preview Guide

1. Review the principle and operation of steam distillation.
2. Learn the property test method of cinnamaldehyde.

1. The glassware are checked for cracks and other defects before installed.
2. Confirm whether the joints are tight before extraction and distillation, and check them again after installation.

1. What are the advantages of steam distillation in extracting natural product?
2. Why is cinnamaldehyde oxidized easily? Please explain it with the chemical equation.

26.3 Extraction of orange-peel oil from orange peel

PPT

Orange-peel oil is a yellow or orange liquid, with a pleasant aroma of orange peel, which is obtained from orange peel by steam distillation or pressing.

The physical properties of are as follows: the relative density is 0.848~0.853 (15℃), the refractive index is 1.473~1.475, and the optical rotation is + 88°~ + 98°. It is easily soluble in ethanol and acetic acid, while it is insoluble in water. Its main component is d-limonene (>90%) and decanal (癸醛). The molecular formula of limonene is $C_{10}H_{16}$ and the structural formula is as follows.

D-limonene belongs to monocyclic monoterpenoid (单萜) with the boiling point of 178℃ and the optical rotation of +125.6°. It is widely distributed in nature, and it be contained in lemon oil, orange-peel oil, peppermint oil, oak oil, turpentine and pine needles. d-limonene is relatively stable, which can be atmospheric distilled without being decomposed. It can be used as a flavor for beverage, food, toothpaste, and soap for its lemon flavor.

Material and Reagents

4~6 fresh orange peels, 20~30ml dichloromethane, and anhydrous sodium sulfate.

Experimental Procedure

In a 500ml round-bottom flask place 4~6 fresh orange peel-cut into several pieces or 30g of chenpi crude powder and distill water running the surface of drug. Assemble the volatile oil extraction device (Figure 4-2), heat and reflux to maintain a stable drip rate, and reflux for 40 minutes.

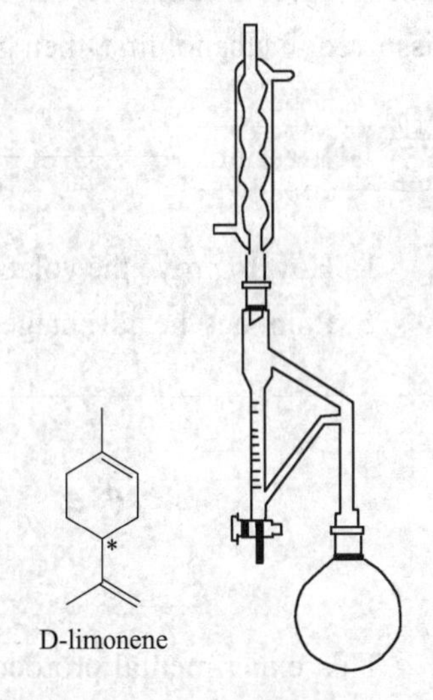

Figure 4-2 Volatile Oil Extractor
图 4-2 挥发油提取装置

Put the distillate into separating funnel, extract the distillate twice using 10ml dichloromethane, and discard the aqueous layer, then dry the extract liquid with anhydrous sodium sulfate.

Filter to discard the desiccant (干燥剂). Recover most of the solvent by distillation on a water bath, then transfer the remaining liquid to a dry test tube, and continue to heat carefully on the water bath at 35℃. Concentrate until the solvent is completely removed, then dry the tube wall. Observe the product color and odor.

Identification: Optical rotation +88°~+98°, refractive index 1.475~1.475. Analyzed by high performance liquid chromatography.

Preview Guide

Preview Requirements

1. Be familiar with the functions and uses of orange-peel oil.
2. Learn the operation and precautions of volatile oil extractor.

Notes

1. The bifurcation in the branch of the extraction device should be parallel to the reference line.
2. Please clean the instrument, especially the calibration part of the receiver carefully. Wash with alcohol and water, followed by a warm mixture of chromic acid and sulfuric acid. Otherwise, the oil layer will adhere to the inner wall and cannot be completely lowered to the bottom of the graduated tube because of the not clean receiver.

Experimental Explanation

1. Using essential oil extractor to extract volatile oil, you can get a preliminary understanding of the content of volatile oil in the medicinal material. The amount of medicinal materials used should be not less than 0.5ml of distilled oil.

2. Volatile oil is insoluble in water, and practically soluble in various organic solvents. They are usually soluble completely in high concentration ethanol. After volatile oils are extracted, they can be dissolved in ethanol first, then get them after the recovery of ethanol.

Questions

1. How to prove the volatile oil in orange-peel oil had been completely distilled off ?
2. Point out the advantages and disadvantages of steam distillation and volatile oil extraction.

26.4 Extraction of star anise volatile oil

PPT

This experimental protocol designed by students, is used to extract the star anise volatile oil from the fruits of *Llicium verum* Hook. f. It has warm property and actions of warming channel for dispelling cold and regulating qi-flowing for relieving pain. Volatile oils in *Llicium verum* Hook. f. has the effects of analgesic, anti-inflammatory, antibacterial. The volatile oil of *Llicium verum* Hook. f. is colorless or light-yellow liquid with anise flavor and volatility. Its relative density is 0.980~0.994 (15℃) and the refractive index is 1.5530~1.5600 (20℃). The main ingredients are anethole (茴香脑)(>85%), methyl chavicol (甲基胡椒酚), safrole (黄樟油精), anisaldehyde (茴香醛), anisic acid (茴香酸), etc. These ingredients are all soluble in dichloromethane, chloroform, ethanol and ether, while it is insoluble in water. Thus, the volatile oil in *Llicium verum* Hook. f. can be extracted by steam distillation.

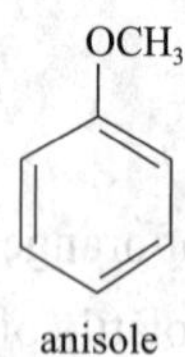

anisole

Instrumental and Reagents

Selection according to the experimental requirements.

Experimental Procedure

1. Access to the literature.
2. Design experimental procedure (include experimental principles, experimental materials,

experimental content, and calculation methods, etc.) and experiment with the teacher's consent.

Preview Guide

Preview Requirements

1. Be familiar with the principle of volatile oil extraction.

2. Seeking information about the compound information and the physical and chemical parameters involved in the experiments.

3. Review the experimental steps of steam distillation and volatile oil extractor.

Notes

1. Volatile oil content measuring devices are generally divided into two types: one is suitable for the determination of volatile oils with a relative density less than 1.0; the other is suitable for the determination of volatile oils with a relative density greater than 1.0.

2. Using essential oil extractor to extract volatile oil, the criterion to judge whether the extraction is complete is that the oil content in the calibration tube of the tester is no longer increased.

Experimental Explanation

1. The determination of volatile oils with a relative density greater than 1.0 is also performed in a tester with a relative density less than 1.0 according to the *Pharmacopoeia of the People's Republic of China*. The method is that before heating, 1ml of dimethylbenzene is added into the measuring device in advance, and then steam distillation is performed to dissolve the volatile oil whose relative density is greater than 1.0 in dimethylbenzene. Because the relative density of dimethylbenzene is 0.8969, generally a mixed solution of volatile oil and dimethylbenzene can float on the water surface. When the amount of the oil layer is read from the scale portion of the tester, the amount of volatile oil is subtracted from the volume of dimethylbenzene added.

2. The contents of volatile oils in some crude drugs are stipulated in the *Pharmacopoeia of the People's Republic of China*, such as the content of volatile oils in *Lllicium verum* Hook. f. must not be less than 4.0%.

Questions

1. Why is the method of steam distillation adopted to distill the volatile oils in *Lllicium verum* Hook. f. ?

2. What simple method can be used to prove that the volatile oil in *Lllicium verum* Hook. f. has been completely distilled off during steam distillation?

实验二十六 挥发油的提取

实验目的

1. 学习中药挥发油的提取原理和方法。
2. 掌握水蒸气蒸馏操作和挥发油提取器的使用。
3. 学会实验设计的思路和方法。

实验原理

挥发油又称精油，是一类在常温下能挥发的、可随水蒸气蒸馏并与水不相混溶的油状液体的总称。大多数挥发油具有芳香气味。挥发油是一类重要的活性成分，主要应用于医药保健品工业、化妆品工业和食品工业等。

挥发油为多种类型成分的混合物，一种挥发油往往含有几十种到一二百种成分，其中以某种或数种成分占较大分量，其基本组成为脂肪族、芳香族和萜类化合物。

大多数挥发油为无色或淡黄色油状透明液体。大多数挥发油比水轻，仅少数比水重（如丁香油、桂皮油等）。挥发油在水中的溶解度很小，易溶于醚、三氯甲烷、石油醚、二硫化碳和脂肪油等有机溶剂中，能完全溶于无水乙醇，在其他浓度的醇中只能溶解一定的量。当挥发油中掺有脂肪油或萜烯类成分时，在一定浓度乙醇中的溶解度就会减少。因此，《中华人民共和国药典》规定用挥发油在醇中的溶解度可以检查挥发油的纯度。

常用于提取挥发油的方法有蒸馏法、溶剂提取法和超临界萃取法，其中水蒸气蒸馏法最为常用。水蒸气蒸馏法是利用混合物沸点低于混合物中任一组分的沸点，即当水十分充足时，挥发油和水将一起在低于100℃的温度下被蒸馏出来。这种提取方法既可提高挥发油的蒸馏速度，也保证了在操作时不会因温度过高而引起原料中易变成分分解而影响油的产率和质量。色谱法是挥发油分析的主要方法，尤其是气相色谱法。

一、牡丹皮中丹皮酚的提取

牡丹皮是毛茛科植物牡丹的根皮，性微寒，味苦，具有清热凉血，活血散瘀之功效。其主要成分为丹皮酚（含量1.9%~2.3%）和丹皮酚苷等，后者在贮存过程中易分解出丹皮酚。除牡丹皮外，中药徐长卿的根中也含有较多的丹皮酚。丹皮酚为具有芳香气味的白色针状结晶，熔点50℃，具有镇痛、镇静、抗菌作用，临床上用于治疗风湿病、牙痛、胃痛、皮肤病及慢性支气管炎、哮喘等症。

丹皮酚的化学名称为2-羟基-4-甲氧基苯乙酮，结构式如下。

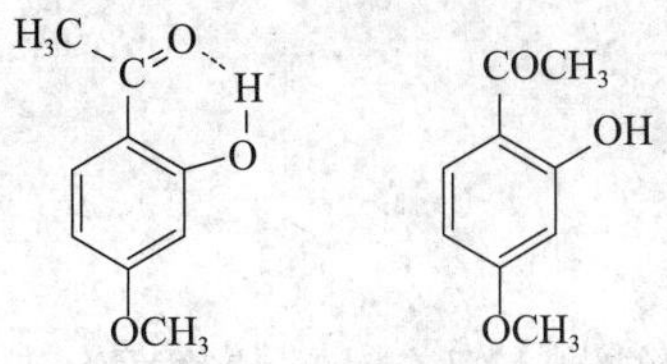

H_3C—C=O…H—O（分子内氢键）　　$COCH_3$, OH

OCH_3　　　　　　　　　　　　OCH_3

丹皮酚分子结构中，处于邻位的羰基和羟基可形成分子内氢键，具有挥发性，能随水蒸气蒸馏。再利用丹皮酚难溶于冷水，易溶于乙醇、乙醚、氯仿、苯等有机溶剂的特性，对其进行纯化。

仪器与试剂

1. 仪器 电子天平、水蒸气蒸馏装置、熔点仪、大烧杯、量筒、减压过滤装置。

2. 试剂 牡丹皮、1% 三氯化铁溶液、食盐、95% 乙醇、碘试液、氢氧化钠溶液。

实验步骤

1. 提取 在 500ml 烧瓶中，加入已粉碎的牡丹皮或徐长卿根 30g，食盐 1g 及适量热水（以能使药材粉末湿润为度），安装水蒸气蒸馏装置，用 250ml 烧杯作接收容器，烧杯内加入食盐 5g，烧杯外用冰水浴冷却。向烧瓶中通入水蒸气进行蒸馏。当馏出液比较清亮、无乳浊现象时，停止蒸馏。将馏出液继续置冰水浴中冷却使固化完全。

2. 分离纯化 馏出液充分放置后，抽滤得到丹皮酚粗品。将结晶用少量 95% 乙醇（不超过 5ml）溶解，再加入大量蒸馏水（约 4 倍的乙醇溶液量）。溶液先呈乳白色，静置后有大量白色针状结晶析出，抽滤出结晶，自然干燥，得丹皮酚纯品。

3. 鉴别

（1）碘仿实验 制备丹皮酚甲醇溶液，碘仿实验应有米黄色沉淀出现。

（2）三氯化铁实验 制备丹皮酚乙醇或甲醇溶液，加三氯化铁检验应显紫红色。

（3）熔点测定 mp 50℃。

预习指导

预习要求

1. 了解丹皮酚的作用与用途。
2. 认识并能写出丹皮酚的分子结构。
3. 熟悉水蒸气蒸馏装置中各部位的名称、作用及操作注意事项。

注意事项

1. 不同季节及不同地域牡丹皮中的丹皮酚含量差别较大，春季及四川省牡丹皮中的丹皮酚含量较高。

2. 在进行水蒸气蒸馏时，理论上需要蒸至馏出液用三氯化铁检验无色为止，即无丹皮酚阳性反应。但如此要花费较长的时间，效率太低。故常蒸馏至馏出液透亮无色为宜。

实验说明

1. 水蒸气蒸馏操作详见实验七。

2. 若在提取过程中得不到白色结晶，只有油珠状物质沉于馏出液下，此时可在馏出液中加入少量丹皮酚晶体（作为晶种），或摩擦瓶壁，即会有大量的白色针状晶体析出。也可加入乙醚振摇萃取三次（30ml、20ml、15ml），合并乙醚提取液，用无水硫酸钠脱水，回收乙醚至少量。静置一夜，即有白色晶体析出。

3. 加入食盐（报道中，食盐加入量为牡丹皮量5%~10%不等），可明显提高丹皮酚的提取率和析出、缩短提取时间。丹皮酚在热水中有一定的溶解度，在提取液中加一定浓度的食盐降低了其在水中的溶解，加速其溶出 / 析出，从而缩短了提取时间，提高提取率。

思考题

1. 结合化学结构，回答丹皮酚为什么具有挥发性？
2. 在进行化学检识（碘仿实验）时，丹皮酚可用乙醇做溶剂吗？为什么？
3. 水杨酸也可用水蒸气蒸馏法提取分离吗？为什么？
4. 根据所学知识，举出你所熟悉的可采用水蒸气蒸馏法提取或分离的实例。

二、从肉桂中分离肉桂醛

肉桂醛，即3-苯基丙烯醛，是一种醛类有机化合物，大量存在于肉桂等植物体内。自然界中天然存在的肉桂醛均为反式结构，其结构式如下：

（结构式：苯环—CH=CH—CHO）

肉桂醛主要用作饮料和食品的增香剂，以及其他产品的调香。肉桂醛纯品为无色或淡黄色液体，相对密度（20℃）1.046~1.052，熔点 −7.5℃，折光率（20℃）1.619~1.623，沸点（常压）253℃。难溶于水，易溶于醇、醚、二氯甲烷等有机溶剂，能随水蒸气挥发。

肉桂醛在强酸性或者强碱性介质中不稳定，在空气中易氧化成肉桂酸。

提取肉桂醛的方法有水蒸气蒸馏法、压榨法和溶剂萃取法。本实验采用水蒸气蒸馏法，提取率为2.0%左右。

仪器与试剂

1. 仪器 50ml圆底烧瓶、蒸馏头、温度计、温度计套管、冷凝管、尾接管、接收瓶、分液漏斗。

2. 试剂 肉桂皮、二氯甲烷、无水硫酸钠、Tollens试剂、Schiff试剂、Br_2/CCl_4溶液。

实验步骤

1. 肉桂醛的提取 取10g肉桂皮，在研钵中研碎，放入50ml圆底烧瓶中，加水20ml，装上冷凝管，加热回流20min。冷却后倒入蒸馏瓶中进行水蒸气蒸馏，收集馏出液8~10ml。将馏出液转移到20ml的分液漏斗中，二氯甲烷萃取三次（$3 \times 2ml$）。弃去水层，合并二氯甲烷于试管中，加入少量无水硫酸钠干燥，20min后，分出萃取液，在通风橱内用水浴加热蒸去二氯甲烷，得肉桂醛。

2. 肉桂醛的性质试验

（1）取提取液1滴于试管中，加入1滴Br_2/CCl_4溶液，观察是否褪色。

（2）取提取液2滴于试管中，加入2滴2, 4–二硝基苯肼试剂，观察是否有黄色沉淀生成。

（3）取提取液1滴于试管中，加入2滴Tollens试剂，水浴加热，观察有无银镜产生。

（4）取提取液1滴于试管中，加入2滴Schiff试剂，振摇，一分钟后，观察是否呈现紫红色，若紫红色不出现，可采用水浴微热2~3min。

预习指导

预习要求

1. 复习水蒸气蒸馏的原理和操作方法。
2. 学习肉桂醛的性质检验方法。

注意事项

1. 安装玻璃仪器前，要检查有无裂缝和其他缺陷。
2. 提取及蒸馏前要确认连接点是否紧密，装好仪器后，再次检查确认。

思考题

1. 水蒸气蒸馏提取天然物质有什么优点？
2. 肉桂醛容易氧化的原因是什么？请用化学方程式解释。

三、从橘皮中分离橘皮油

橘皮油是一种天然香精油，可从柑橘的果皮经水蒸气蒸馏或压榨得到，为黄色或橙色液体，具有橘皮愉快的香味。比重0.848 ~ 0.853（15℃），折光率1.473 ~ 1.475，旋光度+88 ~ +98°，不溶于水，溶于乙醇和冰醋酸，其主要成分是右旋苧烯（含量在90%以上），并含有癸醛等。苧烯的分子式为$C_{10}H_{16}$，结构为

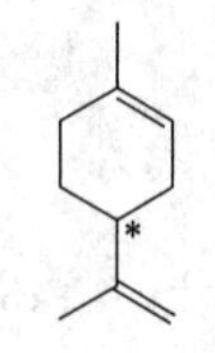

D-limonene

右旋苧烯的沸点 178℃，旋光度 +125.6°，属单环单萜类化合物，广泛存在于自然界中，在柠檬油、橘皮油、薄荷油、橡树油、松节油和松针中都含有。右旋苧烯是一种比较稳定的物质，可在常压下蒸馏而不分解。因具有柠檬的香味，可用做饮料、食品、牙膏、肥皂等的香精。

仪器和试剂

4~6 个鲜橘果皮、20~30ml 二氯甲烷、无水硫酸钠。

实验操作

将 4~6 个鲜橘果皮剪碎或 30 克陈皮，置于一个 500ml 圆底烧瓶中，并加入适量水，没过药材，组装挥发油提取装置（图4-2）。加热回流，维持稳定滴速，回流 40min。

将收集到液体转移至分液漏斗中，每次用 10ml 二氯甲烷萃取，萃取两次，弃去水层，用无水硫酸钠干燥萃取液。

滤弃干燥剂，在水浴上蒸馏回收大部分溶剂，将剩余液体移至一支干燥的试管中，继续在 35℃ 水浴上小心加热，浓缩至完全除净溶剂为止，擦干试管外壁。观察产品颜色、气味如实记录。

鉴定：其旋光度：+88 ~ +98°，折光率：1.475 ~ 1.475，高效液相色谱法。

预 习 指 导

预习要求

1. 熟悉橘皮油的功能和用途。
2. 学习挥发油提取器的操作及注意事项。

注意事项

1. 装置中挥发油提取器的支管分岔处应要与基准线平行。

2. 应注意仔细清洁仪器，尤其是接收器的刻度部分，用酒精和水洗，接着用温热的铬酸硫酸混合液洗。否则，在放油层读数时，由于接收器未清洁干净，油层会黏附于内壁，不能完全下降到刻度管底部。

实验说明

1. 采用挥发油提取器提取挥发油，可以初步了解该药材中挥发油的含量，但所用的药材量应使蒸出的挥发油量不少于 0.5ml 为宜。

2. 挥发油不溶于水，而易溶于各种有机溶剂。在高浓度的乙醇中能全部溶解，所以挥发油提取出来后，可以先把它溶到乙醇中，再回收乙醇即可。

1. 如何判断橘皮中的挥发油已提取完全?
2. 指出水蒸气蒸馏法和挥发油提取器提取挥发油两种方法的优缺点。

四、八角挥发油的提取

本实验方案由学生自己设计从八角果实中提取八角挥发油。八角性温，具有温经散寒，行气止痛的功效，所含的挥发油，具有止痛、消炎、抗菌作用。八角挥发油为无色或淡黄色液体，具有茴香味和挥发性，相对密度为0.980～0.994（15℃），折光率为1.5530～1.5600（20℃）。其主要成分为大茴香醚（占85%以上）、胡椒酚甲醚、黄樟醚、茴香醛、茴香酸等，这些成分均易溶于二氯甲烷、氯仿、乙醇、乙醚等有机溶剂，难溶于水。因此，可采用水蒸气蒸馏法提取。

仪器和试剂

由学生根据实验要求选取。

实验步骤

1. 查阅有关文献。
2. 设计实验方案（包括实验原理、所需实验材料、实验内容、计算方法等），经教师审阅同意后，进行实验。

预 习 指 导

预习要求

1. 熟悉挥发油提取的原理。
2. 查找实验涉及的化合物信息及理化参数。
3. 复习水蒸气蒸馏法和挥发油测定器的实验操作步骤。

注意事项

1. 用挥发油含量测定装置一般分为两种。一种适用于相对密度小于1.0的挥发油测定；另一种适用于测定相对密度大于1.0的挥发油。
2. 用挥发油提取器提取挥发油，以测定器刻度管中的油量不再增加作为判断是否提取完全的标准。

1.《中华人民共和国药典》规定，测定相对密度大于1.0的挥发油，也在相对密度小于1.0的测定器中进行。其方法是在加热前，预先加入1ml二甲苯于测定器内，然后进行水蒸气蒸馏，使蒸出的相对密度大于1.0的挥发油溶于二甲苯中。由于二甲苯的相对密度为0.8969，一般能使挥发油与二甲苯的混合溶液浮于水面。由测定器刻度部分读取油层的量时，扣除加入二甲苯的体积即为挥发油的量。

2.《中国药典》中一些生药规定了挥发油的含量要求。如八角茴香含挥发油不得少于4.0%。

1. 为什么八角挥发油可用水蒸气蒸馏法提取?
2. 在进行水蒸气蒸馏时，用什么简便的方法可以证明八角油已被完全蒸出?

27 Preparation of Nikethamide

Experimental Purpose

1. Master the principle and method of preparing nikethamide from carboxylic acid and amine.
2. Understand the roles of $POCl_3$ and NaOH in the reaction.

Experimental Principle

N, N-diethyl niacinamide also known as Nikethamide is a light color, viscous, oily liquid, or crystalline solid. Melting point 24~26℃, boiling point 296~300℃ (slightly decomposed), relative density d_4^{25}1.058-1.066, refractive index n_D^{20}1.525-1.526. It can be miscible with water, soluble in ethanol, ether acetone and chloroform, etc. It also includes hygroscopicity (吸湿性), mild aroma and bitter taste.

As a central nervous system stimulants, Nikethamide is an essential drug with few side effects. It is widely used in the treatment of central respiratory, circulatory failure, anesthetics and toxication of other central depressants.

Niacin reacts with diethylamine to form a salt, which is condensed to form nicotinyl diethylamine hydrochloride. Nikethamide is obtained by neutralization (中和).

The equations are as follows:

$$C_5H_4N\text{-}COOH + HN(C_2H_5)_2 \longrightarrow C_5H_4N\text{-}COOH\cdot HN(C_2H_5)_2$$

$$C_5H_4N\text{-}COOH\cdot HN(C_2H_5)_2 \xrightarrow{POCl_3} C_5H_4N\text{-}C(=O)N(C_2H_5)_2 + H_3PO_4$$

$$C_5H_4N\text{-}C(=O)N(C_2H_5)_2\cdot HCl + NaOH \longrightarrow C_5H_4N\text{-}C(=O)N(C_2H_5)_2 + NaCl + H_2O$$

Instruments and Reagents

1. Instruments Three-necked round-bottom flask, agitator, separatory funnel, filter apparatus, distillation apparatus, vacuum distillation instruments.

2. Reagents Nicotinic acid, diethylamine, phosphorus oxychloride, 10% potassium permanganate solution, trichloromethane.

Experimental Procedure

In a 100ml dry three-necked round-bottom flask, 12.3g of nicotinic acid and 10.2g of diethylamine are added. Heat with slow agitation and dissolve the powder completely. After cooling below 60°C, 8.4g of phosphorus oxychloride is added dropwise to the flask at such a rate that keep the temperature is below 140°C. Then maintain the temperature at 135°C for 2.5h.

After cooling the mixture to 80°C, 12ml water is slowly added. When the temperature is below 55°C, 20% sodium hydroxide solution is added until the pH is 6~7. Transfer the mixture into a separatory funnel, and remove the aqueous layer. Remove the oil layer to a 100ml conical flask and dilute it with 10ml water, add 3ml 10% potassium permanganate solution and shake well. The oxidized reaction solution is bleached through a funnel coated with activated carbon (about 3g). Wash the filter cake with small amount of water. The washing solution is incorporated into the filtrate, and the pH=7.5 is adjusted with 10% potassium carbonate solution. Transfer the solution to a separating funnel. And extract it with chloroform for 4 times (20ml × 2; 15ml × 2), combine the chloroform, and then wash it with distilled water for 4 times (8ml each time), and dry it with an hydrous sodium carbonate.

The chloroform is removed by air distillation and the yellowish fraction is obtained under reduced pressure distillation at 160~170°C/10~15mmHg, the yield of *N, N*-diethyl nicotinic amide is about 12.5g. Literature value: mp 24~26°C; bp 175°C/25mmHg, 158~159°C/10mmHg, 128~129°C/3mmHg.

Time：5~6 credit hours.

Preview Guide

Preview Requirements

1. Know the function and uses of Nikethamide.
2. Recognize and the molecular structure of niacin.
3. Review the reaction mechanism of carboxylic acid and hydrocarbon amine to form amide.

1. Diethylamine and phosphorus oxychloride should be redistilled before use, and nicotinic acid should be dried below 80°C.

2. If the solid dissolve, heating is left out.

3. Phosphorus oxychloride releases hydrogen chloride upon absorbing moisture. It should be kept under dry conditions and distilled in the fume hood (通风橱).

4. Keep the temperature below 60°C during neutralization to avoid the hydrolysis.

5. Nikethamide is a drug, which should be washed with purified water. If it is washed with tap water, some impurities would be introduced.

Questions

1. What is the role of phosphorus oxychloride in the formation of amide?

2. What would happen if temperature is higher than 60℃ during the neutralization by sodium hydroxide?

3. Why is 10% potassium permanganate (高锰酸钾) solution used to wash the oil layer?

实验二十七 尼可刹米的制备

实验目的

1. 掌握羧酸与胺反应制备尼可刹米的原理及操作方法。
2. 理解三氯氧磷和氢氧化钠在反应中的作用。

实验原理

视频

尼可刹米又称可拉明，其化学名称为 *N, N*- 二乙基烟酰胺，色微黏稠油状液体或结晶固体。熔点 24~26℃，沸点 296~300℃（稍分解），相对密度 1.058~1.066（25/4℃），折光率 1.525~1.526。能与水混溶，易溶于乙醇、乙醚、丙酮或氯仿。有引湿性，微有芳香和苦味。

尼可刹米是一种中枢兴奋药，作为几乎无副作用的基本药物，广泛应用于治疗中枢性呼吸及循环衰竭、麻醉药及其他中枢抑制药的中毒急救。

由烟酸与二乙胺反应成盐，缩合生成烟酰二乙胺盐酸盐，再经中和得到尼可刹米。

反应式：

$$\text{(3-吡啶)COOH} + HN(C_2H_5)_2 \longrightarrow \text{(3-吡啶)COOH}\cdot HN(C_2H_5)_2$$

$$\text{(3-吡啶)COOH}\cdot HN(C_2H_5)_2 \xrightarrow{POCl_3} \text{(3-吡啶)C(=O)}N(C_2H_5)_2 + H_3PO_4$$

$$\text{(3-吡啶)C(=O)}N(C_2H_5)_2\cdot HCl + NaOH \longrightarrow \text{(3-吡啶)C(=O)}N(C_2H_5)_2 + NaCl + H_2O$$

仪器与试剂

1. 仪器 三颈圆底烧瓶、搅拌器、分液漏斗、过滤装置、蒸馏装置、减压蒸馏装置。

2. 试剂 烟酸、二乙胺、三氯氧磷、10% 高锰酸钾溶液、三氯甲烷。

实验步骤

在 100ml 干燥的三颈圆底烧瓶中，加入 12.3g 烟酸、10.2g 二乙胺，开动搅拌，慢慢加热，使固体物全部溶解。溶液冷至 60℃ 以下，慢慢滴加 8.4g 三氯氧磷，控制反应温度不超过 140℃，

滴完后维持 135℃左右反应 2.5h。

将反应混合物冷至 80℃，慢慢加入 12ml 水，待温度降至 55℃后，用 20% 氢氧化钠液中和至 pH=6~7，然后将反应液移至分液漏斗中，弃去水层。将油层移至 100ml 锥形瓶中，加 10ml 水稀释，再加入 10% 高锰酸钾溶液 3ml，摇匀。将氧化后的反应液通过铺有活性炭（约 3g）的漏斗脱色过滤。用少量水洗滤饼，洗涤液合并于滤液中，以 10% 的碳酸钾溶液调 pH=7.5。将溶液转至分液漏斗中，用三氯甲烷提取 4 次（20ml × 2 次，15ml × 2 次），合并三氯甲烷层，用蒸馏水洗涤 4 次（每次 8ml），再用无水碳酸钠干燥。

将三氯甲烷提取液倒入 50ml 烧瓶中，先普通蒸馏除去三氯甲烷，再减压蒸馏收集 160 ~ 170℃ / 10~15mmHg 的馏分，得到微黄液体 12.5g。文献值：熔点 m. p. 24~26℃；沸点 b. p. 175℃/25mmHg，158~159℃/10mmHg，128~129℃/ 3mmHg。

预习指导

预习要求

1. 了解尼可刹米的作用与用途。
2. 认识并能写出烟酸的分子结构。
3. 复习羧酸与烃胺反应生成酰胺的机理。

注意事项

1. 二乙胺及三氯氧磷使用前要重蒸一次。烟酸应在 80℃干燥过。
2. 加料后如固体物已溶，则勿需加热。
3. 三氯氧磷易吸潮，放出氯化氢气体，故应在干燥条件下保存，宜在通风橱内蒸馏。
4. 中和反应温度要控制在 60℃以下，以免水解。
5. 尼可刹米是药物，洗涤时要使用纯净水，若用自来水洗涤则会引入其他杂质，影响产品的质量。

思考题

1. 三氯氧磷在酰胺形成中起什么作用？
2. 用氢氧化钠溶液中和反应液时，若温度高于 60℃会产生什么结果？
3. 用 10% 高锰酸钾洗涤油层的目的是什么？

28 Preparation of Aspirin

Experimental Purpose

1. Master the anhydrous operation and preparation and purification of acetylsalicylic acid.
2. Consolidate the operation of recrystallization.

Experimental Principle

Aspirin (阿司匹林) whose chemica name is acetylsalicylic acid (乙酰水杨酸), is synthesized by salicylic acid (水杨酸)(*o*-hydroxybenzoic acid) and acetic anhydride (乙酸酐). Aspirin is widely used as anti-cold drug, with antipyretic, analgesic, anti-rheumatic (抗风湿) and anti-thrombotic (抗血栓) effects.

$$\text{C}_6\text{H}_4(\text{COOH})(\text{OH}) + (\text{H}_3\text{C}-\text{CO})_2\text{O} \xrightarrow{H^+} \text{C}_6\text{H}_4(\text{COOH})(\text{O}-\text{CO}-\text{CH}_3) + \text{CH}_3\text{COOH}$$

Because of the intra-molecular hydrogen bond (氢键) formed by carboxyl group (羧基) and hydroxyl group (羟基) of salicylic acid, the reaction need be processed at 150 ~ 160℃. If small quantity of sulfuric acid, phosphoric acid (磷酸) or perchloric acid (高氯酸) are added to destroy the hydrogen bond, the reaction temperature can drop to 60 ~ 80℃. Meanwhile, the production of by-product is decreased.

When the acetylsalicylic acid is produced, the condensation reaction is taken place between the products to formulate little amount of polymer (聚合物). Acetylsalicylic acid can react with sodium bicarbonate to form water soluble sodium salt. The polymer cannot resolve in the sodium bicarbonate solution, which can be used to purify the Asprin. Moreover, the aspirin product will also contain the impurity salicylic acid, which is caused by the incomplete acetylation reaction or the hydrolysis (水解) of the product in the separation step. Salicylic acid can be removed by purification and recrystallization.

Instruments and Reagents

1. **Instruments** Erlenmeyer flask, Büchner funnel, thermometer, beaker, glass bottle.
2. **Reagents** Salicylic acid, acetic anhydride, sulfuric acid.

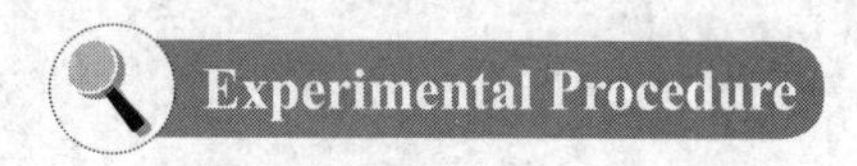

1. Crude product preparation

(1) Approach One In a dry 125ml Erlenmeyer flask, add 2g (0.0145mol) of salicylic acid, 5ml of acetic anhydride[1] and 0.5ml of sulfuric acid (con.) and heat them on water bath to 85-90°C, till the salicylic acid solved completely (you can shake the flask gently if necessary). Then keep heating for 5-10min. After cooling to room temperature (BE SLOWLY), the crystal of acetylsalicylic acid is separate out (if not, you can use glass bar rubbing bottle wall and leave the bottle in the ice water to form crystallization). If the reaction mixture turning mushy, add 50ml of cold distilled water while stirring to decompose excessive acetic anhydride. The mixture is kept cooling in ice water bath to crystallize completely. After filtrating under reduced pressure, the crystal left in the bottle are washed repeatedly by filtrate until all the crystal are collected by the Buchner funnel. All of the crystal is washed by litter distilled water for 3 times and suction to dry as much as possible. The crude products are transferred to the watch glass and dried by atmosphere and weight about 1.8g.

(2) Approach Two Add 2g salicylic acid, 0.1g solid sodium carbonate and 1.8ml acetic anhydride into a dry 50ml Erlenmeyer flask. Put the Erlenmeyer flask into water bath of 85~90°C and keep shaking gently to solve the salicylic acid completely and heat for 10min. Stir the reaction mixture and pour it into a beaker with 30ml distilled water and 0.5ml 10% hydrochloric acid. Cool the beaker in ice water bath for 15min to crystallize completely. Filter the crystal by reduced pressure and wash it by little cold water 2~3 times to yield about 2.0g of the crude product.

2. Purification of crude products-recrystallization Transfer the crude product obtained using approach to a 150ml beaker and add 25ml saturated sodium bicarbonate solution under stirring. Stop string until there is no bubbling (CO_2). The insoluble compound and the byproduct polymer are filtered, and washed with 5~10ml water. The filtrates are combined and poured into a beaker with 4~5ml concentrated hydrochloric acid and 10ml water mixture. After string well, the crystal may separate out. Cool the beaker into ice water bath to crystallize it completely. Filter under reduced pressure, squeeze the filter paper with a clean glass plug, drain the filtrate as much as possible, wash it with a small amount of cold water for 2~3times, and drain the water. The crystal are transferred to the watch glass and dried, about 1.5g.

In order to get purer products, solve half of the above crystal into the minimum amount of ethyl acetate (2~3ml). It is necessary to heat carefully in water bath in dissolution process. If insoluble compounds appear, preheated glass funnel can be taken to filtrate. Cool filtrate to room temperature and Asprin separates out. If no crystal appears, the filtrate can be concentrated in water bath and then put into ice water, or rub the bottle wall with a glass rod. Filter and collect the products, the melting point of which are measured after dry.

Acetylsalicylic acid is a white acicular crystal with a melting point of 132~136°C. This experiment takes about 4 ~ 5 hours.

Preview Guide

Preview Requirements

1. Review the principles of esterification of phenolic hydroxyl groups.

2. Consult the information and physical and chemical parameters of the compounds involved in this experiment, mainly including salicylic acid and acetic anhydride.

3. Analyze the by-products that may be generated during the synthesis process..

4. Review the key points of recrystallization operation.

Notes

1. Acetic anhydride should be freshly evaporated, and 139~140°C fraction should be collected.

2. The reaction temperature should not be too high to exceed 90°C, or the by–products will increase. The Erlenmeyer flask should not be removed from the water bath during the reaction, which will lead to the precipitation of acetylsalicylic acid generated in the reaction and impossible to judge whether the salicylic acid is completely dissolved. If the solid cannot be dissolved after heating for 0.5h, the reaction can be considered complete. Because salicylic acid can be precipitated in solution.

3. After heating on the water bath, it must be cooled slowly and naturally. If the faster cooling speed is, the easier the oil rather than crystal produce.

Experimental Explanation

1. The purpose of acidification is to make acetylsalicylic acid free. The free acetylsalicylic acid has little water solubility and can be separated out.

2. Acetylsalicylic acid is easily decomposed by heat, so the melting point is more difficult to determine and has no fixed value. When measuring the melting point, the heat carrier should be heated to about 120°C, and then put into the sample for measurement.

3. Pure acetylsalicylic acid has no free phenolic hydroxyl groups and does not react with ferric trichloride in color, so it can be used to test whether the synthesis reaction is complete.

Questions

1. What is the purpose of adding concentrated sulfuric acid to prepare aspirin?

2. What are the by-products in the reaction? How to remove?

3. Why should the reaction apparatus be dry and anhydrous? How does the presence of water affect the reaction?

实验二十八　阿司匹林的制备

实验目的

1. 掌握无水操作方法及阿司匹林的制备和纯化。
2. 巩固重结晶的操作方法。

实验原理

阿司匹林（Aspirin），其化学名称为乙酰水杨酸，是由水杨酸（邻羟基苯甲酸）和乙酸酐合成的。阿司匹林是一个广泛使用的治疗感冒的药物，具有解热镇痛、抗风湿、抑制血栓形成等作用。

$$C_6H_4(OH)COOH + (H_3C-CO)_2O \xrightarrow{H^+} C_6H_4(O-CO-CH_3)COOH + CH_3COOH$$

由于水杨酸中的羧基与羟基能形成分子内氢键，反应需加热 150 ~ 160℃。若加入少量的浓硫酸、浓磷酸或高氯酸等来破坏氢键，则反应可降到 60 ~ 80℃ 进行，同时还减少了副产物的生成。

在生成乙酰水杨酸的同时，水杨酸分子之间可以发生缩合反应，生成少量的聚合物。乙酰水杨酸能与碳酸氢钠反应生成水溶性钠盐，而副产物聚合后不能溶于碳酸氢钠，这种性质上的差别可用于阿司匹林的纯化。另外，阿司匹林产物中还会存在杂质水杨酸，这是由于乙酰化反应不完全或由于产物在分离步骤中发生水解造成的，它可以在各步纯化过程和产物的重结晶过程中被除去。

仪器与试剂

1. 仪器　锥形瓶、布氏漏斗、温度计、烧杯、玻璃棒。

2. 试剂　水杨酸、乙酸酐、浓硫酸。

实验步骤

1. 粗产品制备

（1）方法一　在干燥的 125ml 锥形瓶中，加入 2g (0.0145mol) 水杨酸，5ml 乙酸酐和 0.5ml 浓硫酸，旋摇锥形瓶，在 85~90℃ 水浴上加热，使水杨酸全部溶解，并随时振摇，继续在水浴上加热 5~10min。冷至室温（一定要缓慢自然冷却），即有乙酰水杨酸结晶析出（如不结晶，可

用玻璃棒摩擦瓶壁并将反应物置于冰水中冷却使结晶产生）。当反应物呈糊状时，在不断搅拌下加入 50ml 冷的蒸馏水分解过量的乙酸酐，将混合物继续在冰水浴中冷却，使结晶析出完全。减压过滤，用滤液反复淋洗锥形瓶剩余的结晶，直至所有晶体被收集到布氏漏斗。用少量冷水洗涤结晶三次，继续抽滤将溶剂尽量抽干。粗产物转移至表面皿上，在空气中风干，称重，粗产物约 1.8g。

(2) 方法二　在干燥的 50ml 锥形瓶中，依次加入 2g 水杨酸、0.1g 无水碳酸钠和 1.8ml 乙酸酐。将锥形瓶放入 85~90℃ 的水浴中，不断振摇使水杨酸全部溶解，加热 10min，趁热把反应液在不断搅拌下倾入盛有 30ml 冷蒸馏水和 0.5ml 10% 盐酸的烧杯中，然后置于冰水浴中冷却 15min，使结晶析出完全，减压过滤，用少量冷水洗涤结晶 2~3 次，得粗产物约 2.0g。

2. 粗产品纯化——重结晶　将粗产物转移至 150ml 烧杯中，在搅拌下加入 25ml 饱和碳酸氢钠溶液，加完后继续搅拌几分钟，直至无二氧化碳气泡产生。抽滤除去不溶物，副产物聚合物应被滤出，用 5~10ml 水冲洗漏斗，合并滤液，将滤液倾入盛有 4~5ml 浓盐酸和 10ml 水配成溶液的烧杯中，搅拌均匀，即有乙酰水杨酸沉淀析出。将烧杯置于冰浴中冷却，使结晶完全。减压过滤，用洁净的玻塞挤压滤纸，尽量抽去滤液，再用少量冷水洗涤 2~3 次，抽干水分。将结晶移至表面皿上，干燥后约 1.5g。

为了得到更纯的产品，可将上述结晶的一半溶于最少量的乙酸乙酯中（需 2~3ml），溶解时应在水浴上小心地加热。如有不溶物出现，可用预热过的玻璃漏斗趁热过滤。将滤液冷至室温，阿司匹林晶体析出。如不析出结晶，可在水浴上稍加浓缩，并将溶液置于冰水中冷却，或用玻璃棒摩擦瓶壁，抽滤收集产物，干燥后测其熔点。

乙酰水杨酸为白色针状晶体，熔点 132~136℃。本实验需 4~5 小时。

预 习 指 导

预习要求

1. 复习酚羟基酯化反应的原理。
2. 查阅本实验涉及的化合物的信息及理化参数，主要包括水杨酸和乙酸酐。
3. 分析合成过程中可能产生的副产品。
4. 复习结晶操作要点。

注意事项

1. 乙酸酐应是新蒸的，收集 139~140℃ 馏分。

2. 反应温度不宜过高超过 90℃，否则副产物增多。反应过程不要将锥形瓶移出水浴，这样会导致反应生成的乙酰水杨酸析出，从而无法判断水杨酸是否完全溶解。如果加热 0.5 小时依然有固体无法溶解，可认为反应完全。因水杨酸溶解后可在溶液中析出。

3. 水浴上加热后一定要缓慢自然冷却，若冷却速度过快，容易出现油状物而不是晶体。

1. 酸化的目的是使乙酰水杨酸游离，游离的乙酰水杨酸水溶性小，可析出。

2. 乙酰水杨酸易受热分解，因此熔点较难测定，无定值。测定熔点时，应先将热载体加热至

120℃左右，然后放入样品测定。

3. 纯的乙酰水杨酸分子中无游离的酚羟基，不与三氯化铁有颜色反应，因此可用来检验合成反应是否完全。

思考题

1. 制备阿司匹林时，加入浓硫酸的目的何在？
2. 反应中有哪些副产物？如何除去？
3. 反应仪器为什么要干燥无水？水的存在对反应有何影响？

PPT

29 Reduction Reaction of Camphor

Experimental Purpose

1. Learn the principle and operation of reducing camphor (樟脑) with $NaBH_4$.
2. Know the application of thin layer chromatography in synthetic reaction.

Experimental Principle

Camphor can be reduced with sodium borohydride to generate two diastereomers (非对映异构体), borneol (冰片) and isoborneol (异冰片). Due to the high stereoselectivity, the main product is isoborneol. There are different physical properties and different polarities between borneol and isoborneol.

Reaction formula as follows:

camphor $\xrightarrow{NaBH_4}$ borneol + isoborneol

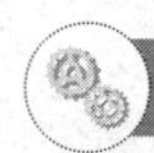

Instruments and Reagents

1. Instruments 25ml Erlenmeyer flask (锥形瓶), 100ml Erlenmeyer flask, reflux condenser, TLC plate, chromatography cylinder, beaker.

2. Reagents camphor, $NaBH_4$, methanol, ethanol, anhydrous Na_2SO_4 or anhydrous $MgSO_4$.

Experimental Procedure

Dissolve 1g camphor in 10ml methanol in a 25ml Erlenmeyer flask and add 0.6g sodium borohydride in batches carefully at room temperature and shake the conical flask while adding. The reaction temperature can be controlled by ice bath if necessary. When all the sodium borohydride is added, the reaction mixture is heated to reflux until the sodium borohydride disappears. Cool the reaction mixture to room temperature, stir and pour the reaction solution into 20g ice water. After all the ice melts, filter and collect the white solid. Wash the white solid with cold water 3 times and dry it in the air.

The solid is transferred to a 100ml Erlenmeyer flask and 15ml of ethyl ether is added to dissolve the solid and then add 2g anhydrous Na_2SO_4 or $MgSO_4$. After drying for 5 minutes, transfer the solution

(desiccant is removed) to a pre-weighed beaker or conical flask. Remove the solvent in a fume hood to generate a white solid, which is further recrystallized with anhydrous ethanol. The products is about 0.6g, mp 212°C, which is identified by thin layer chromatography. The products can also be identified by infrared spectrometry (红外光谱) with potassium bromide.

Preview Guide

Preview Requirements

1. Review the reaction principle and application scope of $NaBH_4$.
2. Review the stereochemical (立体化学) reaction of reducing carbonyl group with $NaBH_4$.

Notes

1. $NaBH_4$ is easily decomposed to release hydrogen after absorbing moisture, so the reagent needs to be stored in a dryer after opening.

2. Ethyl ether is flammable and should be kept away from fire.

Experimental Explanation

1. The preparation method of the thin layer plate is as follows. Take 5g silica gel and 13ml 0.5%~1% sodium carboxymethyl cellulose (羧甲基纤维素钠) solution and mix them well in the mortar; spread it on the clean and dry glass sheet and thickness is about 0.25mm. After drying at room temperature, activate it in an oven at 110°C for half an hour and then take it out. After cooling, put it into dryer for later use.

2. Product identification. A 5 × 15cm TLC plate is taken and solutions of borneol, isoborneol, camphor and reduction products of camphor in ethyl ether are used as samples for analysis of thin layer chromatography. When the eluent reaches the solvent front, take the thin layer plate out. When there is a little eluent left on the thin layer plate, cover it with another glass plate with the same size and evenly coated with concentrated sulfuric acid and then the color reaction occurs. R_f values of four samples can be compared to prove that camphor has been reduced to borneol and isoborneol.

3. Dichloromethane-benzene (2 : 1, v/v) is used as the developer.

Questions

1. What should be paid attention to when determining the melting point of product?

2. In addition to thin layer chromatography, what other methods can be used to distinguish and identify borneol and isoborneol?

3. What is the main product when 2-norbornanone is reduced by $NaBH_4$?

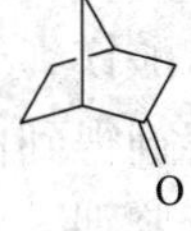

2-norbornanoe

实验二十九 樟脑的还原反应

实验目的

1. 学习用 $NaBH_4$ 还原樟脑的原理及操作方法。
2. 了解薄层层析在合成反应中的应用。

实验原理

用硼氢化钠还原樟脑得到冰片和异冰片两种非对映异构体。由于立体选择性较高，所得产物以异冰片为主。冰片和异冰片具有不同的物理性质，两者极性不同。

反应式：

樟脑 $\xrightarrow{NaBH_4}$ 冰片 + 异冰片

仪器与试剂

1. 仪器 25ml 锥形瓶、100ml 锥形瓶、回流冷凝管、薄层板、层析缸、烧杯。

2. 试剂 樟脑、硼氢化钠、甲醇、乙醚、无水硫酸钠或无水硫酸镁。

实验步骤

在 25ml 锥形瓶中将 1g 樟脑溶于 10ml 甲醇，室温下小心分批加入 0.6g 硼氢化钠，边加边振摇。必要时可用冰水浴控制反应温度。当所有硼氢化钠加完后，将反应混合物加热回流至硼氢化钠消失。冷却到室温，搅拌下将反应液倒入盛有 20g 冰水的烧杯中，待冰全部融化后，抽滤收集白色固体，用冷水洗涤固体 3 次，晾干。

将固体转移至 100ml 的锥形瓶中，加入 15ml 乙醚溶解固体，随后加入 2g 无水硫酸钠或无水硫酸镁，干燥 5 分钟后除去干燥剂，将溶液转移至预先称好的烧杯或锥形瓶中。在通风橱中移除溶剂得到白色固体，并用无水乙醇重结晶，产量约为 0.6g，mp 212℃，所得产物用薄层色谱法进行鉴别，也可用溴化钾压片做产物的红外光谱。

预习指导

预习要求

1. 复习还原剂 $NaBH_4$ 的反应原理及应用范围。
2. 复习 $NaBH_4$ 还原羰基的立体化学反应。

注意事项

1. $NaBH_4$ 吸水后易变质，放出氢气，故开封后的试剂需置于干燥器内保存。
2. 乙醚属易燃物，使用时应远离明火。

实验说明

1. 薄层板的制法 取5g硅胶与13ml 0.5%~1%的羧甲基纤维素钠水溶液，在研钵中调匀，铺在清洁干燥的玻璃片上，厚度约0.25mm。室温晾干后，在110℃烘箱内活化半小时，取出放冷后置干燥器内备用。

2. 产物的鉴别 取一片5×15cm的薄层板，分别用冰片、异冰片、樟脑和樟脑还原产物的乙醚溶液点样，置于层析缸中展开。取出层析板，待薄层上尚残留少许展开剂时，立即用另一块与薄层板同样大小并均匀地涂上浓硫酸的玻璃板覆盖在薄层板上，即可显色。将4个点的 R_f 值对比，证明樟脑已被还原成冰片和异冰片，

3. 以二氯甲烷 - 苯（2∶1, v/v）为展开剂。

思考题

1. 测产物熔点时应注意什么？
2. 除了薄层色谱法之外，鉴别冰片和异冰片的方法还有哪些？
3. 若如下原冰片酮用 $NaBH_4$ 还原时，预计得到的主要产物是什么？

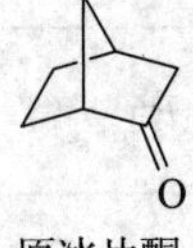

原冰片酮

Appendix ¦ 附录

Name	Abbreviation	Melting Point (℃)	Boiling Point (℃)	Flashing Point (℃)	Relative Density d_4^{20}
acetic anhydride 乙酸酐	Ac_2O	–73	139	49	1.082
acetanilide 乙酰苯胺		114.3	305	173.9	1.2190
aniline 苯胺		–6.3	184.4	70	1.02
anhydrous ether 无水乙醚		–116.3	34.6	–45	0.71
adipic acid 己二酸		152	337	196	1.360
1-bromobutane 正溴丁烷		–112.4	101.6	23	1.30
benzoic acid 苯甲酸		122.13	249	121-123	1.2659
bromobenzene 溴苯		–30.7	156.2	51	1.50
cyclohexene 环已烯		–104	83	–24	0.8102
cyclohexanol 环已醇		25.5	161	67.78	0.962
cinnamic acid 肉桂酸		133			1.25
cinnamaldehyde 肉桂醛		–7.5	253	71.1	1.046~1.052
caffeine 咖啡因		238℃			
d-limonene 右旋苧烯			178℃		0.8402
dichloromethane 二氯甲烷		–97℃	39.8℃	30℃	1.33
diethyl ether 乙醚		–116.2	34.6		0.71
ethanol 乙醇		–114.1	78.3	12	0.79

(Continued)

Name	Abbreviation	Melting Point (℃)	Boiling Point (℃)	Flashing Point (℃)	Relative Density d_4^{20}
ethyl benzoate 苯甲酸乙酯		–34.6	213	93	1.05
ethyl acetate 乙酸乙酯	EtOAc	–84℃	77℃	–4℃	0.902
2-furanmethanol 2- 呋喃甲醇			171		1.1269
2-furoic acid 2- 呋喃甲酸		133~134	232.14	94.195	1.322
glacial acetic acid 冰醋酸		16.5	118.1	39	1.05
2-hydroxy-1, 2-diphenylethan-1-one 安息香		134–136	343.0	154.8±14.9	1.2
N, N-dimethylformamide *N, N*- 二甲基甲酰胺	DMF	–61	153	58	0.95
N, N-dimethylaniline *N, N*- 二甲基苯胺		2.5	193.1		0.96
2-nitro-1, 3-benzenediol 2- 硝基 -1, 3- 苯二酚		84–85	78.4		0.7893
n-butanol 正丁醇		–88.9	117.2	35	0.81
resorcinol 间苯二酚		109~111	281		1.285
urea 尿素		135			1.330

Reference 参考文献

1. 赵骏，杨武德 . 有机化学实验［M］. 2 版 . 北京：中国医药科技出版社，2018.

2. 薛思佳，季萍，Larry Olson. 有机化学实验英汉双语版［M］. 3 版 . 北京：科学出版社，2019.

3. 陆涛，陈继俊 . 有机化学实验与指导［M］. 北京：中国医药科技出版社，2003.

4. Brian S. Furniss, Antony J. Hannaford, Peter W. G. Smith, Austin R. Tatchell. Vogel's Textbook of practical organic chemistry［M］. 5th ed. New York: Longman Scientific & Technical, John Wiley & Sons, 1989.

5. Wang Mei, Wang YanHua, Gao ZhanXian. Organic Chemistry Experiments［M］. 北京：高等教育出版社，2011.

6. 冯文芳 . 有机化学实验（双语）［M］. 武汉：华中科技大学出版社，2014.

7. 梁敬钰 . 天然药物化学实验与指导［M］. 2 版 . 北京：中国医药科技出版社，2010.

8. 袁华，尹传奇 . 有机化学实验（双语版）［M］. 北京：化学工业出版社，2008.

9. 李晓飞，杨静 . 基础化学实验［M］. 北京：中国中医药出版社，2017.

10. 高增平 .《中药化学》（英文版）［M］. 北京：中国中医药出版社 . 2014.